I0470669

U.S. Department
of Transportation
**National Highway
Traffic Safety
Administration**

www.nhtsa.gov

DOT HS 811 206

December 2009

Lives Saved Calculations for Seat Belts and Frontal Air Bags

1. Report No. DOT HS 811 206	2. Government Accession No.	3. Recipient's Catalog No.
4. Title and Subtitle Lives Saved Calculations for Seat Belts and Frontal Air Bags		5. Report Date December 2009
		6. Performing Organization Code NVS-421
7. Author(s) Glassbrenner, Donna, Ph.D., and Starnes, Marc		8. Performing Organization Report No.
9. Performing Organization Name and Address Mathematical Analysis Division, National Center for Statistics and Analysis National Highway Traffic Safety Administration NVS-421, 1200 New Jersey Avenue SE. Washington, DC 20590		10. Work Unit No. (TRAIS)
		11. Contract or Grant No.
12. Sponsoring Agency Name and Address Mathematical Analysis Division, National Center for Statistics and Analysis National Highway Traffic Safety Administration NVS-421, 1200 New Jersey Avenue SE. Washington, DC 20590		13. Type of Report and Period Covered NHTSA Technical Report
		14. Sponsoring Agency Code
15. Supplementary Notes		

Abstract

One of the ways the National Highway Traffic Safety Administration quantifies the benefits of seat belts and frontal air bags is to estimate the number of passenger vehicle occupants whose lives were saved by these protective devices, and the lives savable if more passenger vehicle occupants had buckled up. Passenger vehicles include passenger cars, sport utility vehicles, pickup trucks, and vans.

This report details how NHTSA produces these lives saved estimates for seat belts and frontal air bags. The methodology is described in detail, including the use of effectiveness ratings for seat belts and frontal air bags. Seat belt effectiveness ratings vary according to the seat belt type (i.e., 3-point belt versus 2-point lap belt), vehicle type, occupant seating position, and occupant age. Frontal air bag effectiveness ratings are consistent for all passenger vehicles. The interactions of the effectiveness of seat belts and the effectiveness of frontal air bags are discussed in this report. The effectiveness of side air bags, child safety seats, motorcycle helmets, and other safety devices are discussed in other NHTSA reports, and are not quantified within this report.

When calculating lives saved by seat belts and air bags, it is important to remember that these calculations are estimates based on a methodology that uses the tremendous amount of knowledge that National Highway Traffic Safety Administration has gained. This knowledge has been expanded through the analysis of decades of data on the roles played by seat belts and air bags in fatal crashes.

17. Key Words lives saved, effectiveness, seat belts, air bags, potential lives saved	18. Distribution Statement Document is available to the public from the National Technical Information Service www.ntis.gov		
19. Security Classif. (of this report) Unclassified	20. Security Classif. (of this page) Unclassified	21. No. of Pages 68	22. Price

Form DOT F 1700.7 (8-72)   Reproduction of completed page authorized

Table of Contents

Tables

Figures

1. Summary

This report examines the calculation of three prominent NHTSA estimates:

<u>Lives saved by seat belts</u> - the estimated number of passenger vehicle occupants who were saved in any given year by seat belts;
<u>Lives saved by frontal air bags</u> - the estimated number of passenger vehicle occupants who were saved in any given year by frontal air bags; and
<u>Potential lives saved by seat belts</u> - the estimated number of passenger vehicle occupants who would have been saved if more had chosen to use their seat belts.

These estimates are published annually by NHTSA, and are cited in reports and speeches to substantiate the benefits of occupant protection devices and to underscore the importance of raising seat belt use as quickly as possible. For estimates of the number of lives saved in 2007, refer to Starnes, 2008.

In this report, we explain concepts such as potential fatalities, effectiveness ratings, and attribution methods, first in the context of a single device, then for two devices, and finally for seat belts and frontal air bags. The one-device descriptions explain the basic ideas and formulas associated with the concept; the two-device sections explain how complications that arise with multiple devices are treated; and the sections on seat belts and frontal air bags apply the two-device formulas and explain additional considerations that arise for these particular devices. Note: This report focuses exclusively on seat belts and frontal air bags, and does not estimate the number of lives saved by other types of air bags, such as torso air bags or head-protection air bags, which are two types of side air bags.

This report also discusses how the estimated number of lives saved by some device, such as a seat belt, depends on which other devices one simultaneously calculates benefits for. For instance, NHTSA's estimates of the lives saved by seat belts and frontal air bags are computed in a calculation that estimates the lives saved by these two restraints, but does not estimate the lives saved by any other devices that also provide protection in crashes, such as reinforced passenger compartments. Incorporating additional safety devices would result in equally valid calculations that produce different estimates of the lives saved by seat belts and air bags. It is important to note that the calculation that only estimates the lives saved by seat belts and frontal air bags does not ignore the life-saving qualities of reinforced passenger compartments. It simply does not estimate the compartments' benefits. In the appendix, we investigate the ramifications of the choice of devices for which to simultaneously calculate benefits.

We discuss the choice of the particular fatality cases to use in the calculations, and how to treat these cases when crucial information is missing. Ratings for the effectiveness of seat belts and air bags are incorporated, as well as a model for seat belt use used in the calculation of the lives savable at higher seat belt use rates.

2. Introduction

Seat belts and frontal air bags are among the most important safety devices in society today, together saving thousands of lives each year. The National Highway Traffic Safety Administration quantifies the benefits of these devices by estimating the numbers of people saved by each device, the number who would have lived if more occupants had buckled up, and the corresponding expressions of the  savings and loss in financial terms. This information is then used in NHTSA literature and speeches to promote belt use, and is used to perform cost-benefit analyses of proposed regulations concerning belts and bags.

In 2007, an estimated 15,147 lives were saved by seat belts, and 2,788 lives were saved by air bags. If seat belt use increased to 100 percent, then an additional 5,024 lives would have been saved (Starnes, 2008). When these seat belt numbers are added together to equal 20,171 (= 15,147 + 5,024), this estimate is often referred to as "potential lives saved" or "lives savable."

This report concerns the calculation of these estimates of lives saved by belts and air bags, and the lives savable if more occupants had used their belts. For information on the calculations of cost savings see Blincoe et al., 2002.

NHTSA has estimated the lives saved by seat belts since 1975 and those by frontal air bags since 1985, when air bags started appearing in appreciable numbers of vehicles. These are computed annually by NHTSA's National Center for Statistics and Analysis (NCSA), and are published in NCSA's Occupant Protection Fact Sheets, such as NCSA's *Traffic Safety Facts 2007 – Occupant Protection* (undated). Specifically, NCSA computes the following items on an annual basis:

- the number of passenger vehicle occupants age 5 and older saved by seat belts;
- the number of passenger vehicle occupants 13 and older saved by frontal air bags; and
- the number of passenger vehicle occupants 5 and older who would have lived if belt use had been higher than it was in the given year.

Passenger vehicles comprise passenger cars, vans, SUVs, and pickup trucks. Note that the estimates of savable lives (or potential lives saved) (i.e., the third bullet, above) refer to belt use in the front seat during daytime, rather than in all seating locations and during all times of day. The "front-seat daytime" restriction is necessitated by practical reasons. (The calculation uses data from an observational survey of belt use, and it can be difficult to observe use at night or in the rear seat.) However we would expect that if front-seat daytime use increases, then use would increase to some, perhaps lesser, amount in other seats and at other times of day. In fact, the calculation of potential lives saved will use a model (see Chapter 7) that will reflect this expectation. For brevity, when describing potential lives saved (also referred to as "lives savable"), the words "front-seat daytime" are frequently dropped, with the third bullet just described as the (potential) lives saved if belt use had been higher.

None of the lives saved estimates in this report reflect the relatively small numbers of large-truck occupants saved by belts, children under 5 saved by belts, children under 13 saved by frontal air bags, and occupants saved by side air bags. NHTSA recommends that children under 13 not be in front of an air bag, unless no other seating position is available.

NCSA calculates the numbers of children under 5 saved by belts and child restraints, and publishes them in the same Occupant Protection Fact Sheets. See Starnes (2005) to see how these are calculated. The changes we present in this report do not affect current and previous child computations because the air bag ratings used by NHTSA for children are effectively neutral.

Note that the third bullet above comprises several estimates, one for each hypothesized higher belt use. Our calculation will allow any hypothesized use, but NHTSA typically publishes the savings at the following hypothesized rates:

- belt use that is 1 percentage point higher than the rate in a given year,
- 90 percent use, and
- 100 percent use.

The numbers of lives saved and potential lives saved are not derived by a case-by-case examination of serious crashes. Even if such crashes were reported to NHTSA, it would be exceedingly difficult and highly subjective to decide in a given crash whether an occupant did not die because they used a belt (or bag), or whether they died because they did not use a restraint. Instead the calculations are based on the numbers of fatalities, the restraints they used, and the effectiveness of these restraints for preventing fatality. We explain the calculations in detail in this report.

This report incorporates NHTSA's most recent ratings (see Table 2, page 18) for the effectiveness of seat belts and air bags, which are published in Kahane (2000) and Morgan (1999). These devices are periodically re-rated by NHTSA for effectiveness, in order to reflect changes in the characteristics of vehicles, crashes, and motorists.

Also included in this report is a model for belt use used when calculating estimates of the potential lives saved had higher belt use been achieved in the United States (see Table 13). The model is used to predict the belt use among potential fatalities (those who would have died if they had been unbelted and had not had an air bag) as a function of the observed front outboard belt use in daytime (not limited to potential fatal crashes). This model is from Wang and Blincoe (2003).

2.1 Terminology

In the remainder of this report, the term *air bag* will always mean "frontal air bag," although we occasionally write *frontal air bag* for emphasis. The term *vehicle* will refer to passenger vehicles.

Lives saved by belts will mean "passenger vehicle occupants of age 5 and older saved by belts," and *lives saved by bags* will mean "passenger vehicle occupants 13 and older saved by frontal air bags."

In the context of potential lives saved, *belt use* will mean "front outboard belt use during daytime." *Lives savable if belt use had been x percent* will mean "passenger vehicle occupants over 4 years old who would have been saved if (daytime front outboard) belt use had been x percent." We shall use the terms "potential lives saved," "lives savable," and "lives potentially saved" interchangeably to refer to this quantity.

Finally, *LTV* will denote a light truck or van, that is, a van, sport utility vehicle, or pickup truck.

3. Fatalities Used in the Calculations

Lives saved and potential lives saved are calculated from a list of the numbers of fatalities in a given year, broken out by the various restraint systems in the vehicle and that were used. This section describes how this list, which we call the *fatality counts*, is produced. The counts will comprise any person 5 and older who died within 30 days of being in a crash in a specified year, who was in a passenger vehicle in a location where there was a belt (e.g., not in the bed of a pickup).

3.1 Data Source

NHTSA compiles a census of all motor vehicle fatalities in the United States from police reports called the Fatality Analysis Reporting System (FARS). A fatal crash is defined as a police-reported crash involving a motor vehicle in transport on a public road, street, or highway in which at least one person, called a fatality, died within 30 days of the crash. This section describes the information we extract from FARS for the fatality counts. It also describes the treatment of unknowns (e.g., when we do not know the age of an occupant who died).

The fatality count is produced using the following FARS variables: body type (BODY_TYP), make and model (MAKE_MOD), VIN-derived model (VINA_MOD), vehicle identification number (VIN), towed trailing unit (TOW_VEH), VIN-derived truck series (SER_TR), model year (MOD_YEAR), person type (PER_TYP), age (AGE), seating position (SEAT_POS), restraint use (REST_USE), and injury severity (INJ_SEV).

Note that multiple model information (MAKE_MOD, VIN, VIN model, VIN truck series) is used. These are used by Kahane (2000) to determine the type of seat belts in a vehicle and whether the vehicle has air bags.

3.2 Inclusion Criteria

The fatality counts reflect the people in FARS in a given data year who meet all of the following criteria. These criteria identify passenger vehicle occupants 5 and older who died from motor vehicle crashes and had access to belts where they were seated, regardless of whether the belts were used.

Inclusion Criteria
1. The person was an occupant.
2. The person died within 30 days of the crash.
3. Either (a) the person was over age 4 at the time of the crash, or (b) age was unknown and the child was not in a child safety seat.

4. The vehicle that the person occupied was a passenger vehicle.
5. The vehicle was known to have been manufactured after 1967, or its model year was unknown.
6. At the time of the crash, the person was located in (a) a position in the first four rows of seats that was not coded in FARS as being an "other" position, (b) an enclosed passenger area in a 15-passenger van, or (c) an unknown seating position.

Note that these criteria might include some people and not others in the same crash (e.g., if one person was in the bed of a pickup and the other was the driver). Note with regard to Criterion 1 that all people in FARS are categorized as either occupants or non-occupants.

In Criterion 4, "passenger vehicle" is defined as in FARS. (See Tessmer [2002] for the specific definition.) This includes passenger cars, vans, SUVs, and pickup trucks whose gross vehicle weight rating (GVWR) does not exceed 10,000 pounds. (The GVWR is a rating determined by the manufacturer that indicates how heavily the vehicle may be safely loaded.) The class of passenger vehicles also includes vehicles known to be some kind of truck and not known to have been towing a trailing unit. It excludes vehicles for which we have no information on the body type.

We exclude vehicles manufactured prior to 1968. Many of these vehicles were not originally equipped with belts. Those that are registered and are not antique vehicles may be required by State laws to be retrofitted with belts, but it is not clear that retrofitted belts will be as effective as those originally installed. Some of these vehicles are antiques that are not required to be retrofitted.

Criterion 6 identifies seating positions that are likely to have a seat belt. We include those positions that obviously have belts (driver, right side of second seat, etc.). We exclude those that obviously do not have belts (the bed of a pickup truck, the exterior of the vehicle, etc.). Occupants in an enclosed passenger or cargo area (code 51 of SEAT_POS) in a 15-passenger van are likely to have been in the fifth-row seat, and so are included. Occupants coded as being in an "other" position in the first through fourth row seats (codes 18, 28, 38, and 48 of SEAT_POS) are likely to have been sitting on someone's lap or on the floor, so are excluded. Occupants in rows 1 to 4 with unknown positions were likely ejected from the vehicle in the crash, and so usually had belt access. We take all occupants whose positions were entirely unknown (code 99 of SEAT_POS) because 99 percent of cases in FARS with known seating positions have belt access.

3.3 Restraint Configurations

NHTSA's effectiveness ratings of belts and bags in Kahane (2000), Morgan (1999), and Report to Congress (2001) are basically specified in terms of the six coordinates in Table 1, which are referred to as the *restraint configurations*.

The coordinate "air bag in seating position?" only refers to frontal air bags, which are only recommended for people over age 12. Consequently this coordinate takes the value "yes" if the occupant is the driver or right-front passenger over 12 years old and there is a driver's (respectively, passenger) frontal air bag, "no" if the occupant is the driver or right-front passenger over 12 years old and there is no driver's (respectively, passenger) frontal air bag, and "NA" if the occupant is in another seating position or is under the age of 13. In the lives saved computation, "NA" is treated as "no," so one could think of this coordinate as only having yes-no values.

For instance, a 15-year-old in the right-front passenger seat of a car using a 3-point belt with no air bag would have the restraint configuration i=(passenger car, right-front passenger, 3-point, yes, no, 13 or older). The same motorist in a rear outboard seat having, but not using a lap belt would have the configuration i=(passenger car, rear outboard, lap, no, NA, 13 or older). Note that restraint configurations contain information on the vehicle and occupant as well as the restraint.

Three–point belts are the manual lap/shoulder belts in today's vehicles and the automatic lap/shoulder belts that appeared primarily in vehicles made by General Motors.

Table 1: Components of Restraint Configurations	
Coordinate	**Values**
vehicle type	passenger car, LTV
seating position	driver, right-front passenger, front-center, rear-outboard, rear-center
belt type	3-point, 2-point, lap/shoulder, lap
belt used?	yes, no
air bag in seating position?	yes, no, NA
age	5 to 12, 13 or older

Two-point belts consist of either (a) an automatic shoulder belt combined with a manual lap belt, a configuration that appeared in some passenger cars in the 1980s and into the 1990s, or (b) an automatic or manual shoulder belt together with a knee bolster under the dashboard, a less common configuration.

Lap/shoulder belts consist of separate lap and shoulder belts. These appeared in some pre-1974 vehicles, and most did not have retractors. We refer to such belts in this paper as "lap/shoulder" However the terms "lap/shoulder" and "3-point" have been used interchangeably by NHTSA in other documents.

Determining the Restraint Configurations
The variables comprising the restraint configurations are created as follows.

Vehicle type   Assign "Passenger car" and "LTV" using standard FARS coding from the body type and towed vehicle variables. These procedures are documented in Tessmer (2008).

Seating position   If SEAT_POS = 51 (enclosed passenger or cargo area) assign "rear outboard." These are 15-passenger vans (since we have excluded the other cases with this seating position). Assign drivers in unknown seating positions (i.e., PER_TYP = 1 and SEAT_POS = 99) to "driver." Assign nondrivers in unknown seating positions (i.e., PER_TYP = 2 or 9, and SEAT_POS = 19, 29, 39, 49, or 99) to "right-front passenger." Assign all other cases in the obvious way.

Age 5-12	Assign "yes" if the age is known to be in this range. (Unknown ages are assumed to be over 12.)
Belt used?	Assign in the obvious way among the cases where belt use was known and distribute the unknowns.
Belt type	Assign as done in Kahane (2000). In the cases for which Kahane (2000) cannot determine a belt type, assign "3-point." This will include all cases with unknown VINs and passengers whose seating position was unknown.
Air bag in seating position?	Assign as in Kahane (2000). Recall that in this report, "air bag" means "frontal air bag." Air bags are required in all passenger cars manufactured after 1998, and in light trucks and vans manufactured after 1999. In the cases for which Kahane (2000) cannot determine air bag presence, assign "yes" if the FARS variable AIR_BAG indicates that there was a bag and "no" otherwise.

We assign the belt type to be "3-point" when we cannot tell what the belt type is, because 3-point belts are by far the most common type in FARS.

In vehicles for which air bags were optional equipment to the consumers, the VIN sometimes does not indicate whether the consumers chose the option. In this case, we effectively assume that the consumer chose the option if AIR_BAG indicates there was a bag and did not otherwise. This will usually result in assigning "No air bag" because the FARS variable AIR_BAG, which comes from police reports, has many missing values. Many State crash report forms do not collect this information.

4. Effectiveness Ratings and Potential Fatalities

The performance of a safety device is measured by its effectiveness rating. Seat belts and air bags are rated for their effectiveness in preventing fatalities and for reducing the severity of injuries in crashes. This report is only concerned with preventing fatalities, although much of its material would pertain to injury reduction as well.

4.1   For a Sole Safety Device

We first explain the relatively simple case of the effectiveness of a solitary safety device. Section 4.2 presents the more complex scenario of multiple devices designed to protect occupants in a common setting, such as belts and bags protecting people in crashes.

4.1.1 Devices and Settings

In general, to estimate lives saved, one must specify a potentially life-threatening situation, called the setting, and one or more factors, called devices, that affect survival. In this report, the setting is the crash of a passenger vehicle and the devices are seat belts and air bags.

In the "single-device scenario" (or "sole-device scenario") there is only one device. We use this simplified situation to introduce the basic concepts of benefits calculations, including effectiveness ratings, potential fatalities, lives saved, and potential lives saved (or savable lives). We then expand these concepts to the more complex two-device scenario used in our calculation for seat belts and air bags.

Figure 1: An Example of a Setting and Device

The Setting
Motorcycle Crashes

The Device
Motorcycle Helmets

4.1.2 Potential Fatalities

In general, when a single device A protects people in a certain setting, then Device A is rated on the hypothetical population of all people in potential instances of that setting who would die without A. This population is said to consist of the *potential fatalities*. For instance, in the single device scenario of motorcycle helmets protecting people in the setting of motorcycle crashes, the potential fatalities would be the motorcyclists in crashes severe enough that they would die without a helmet.

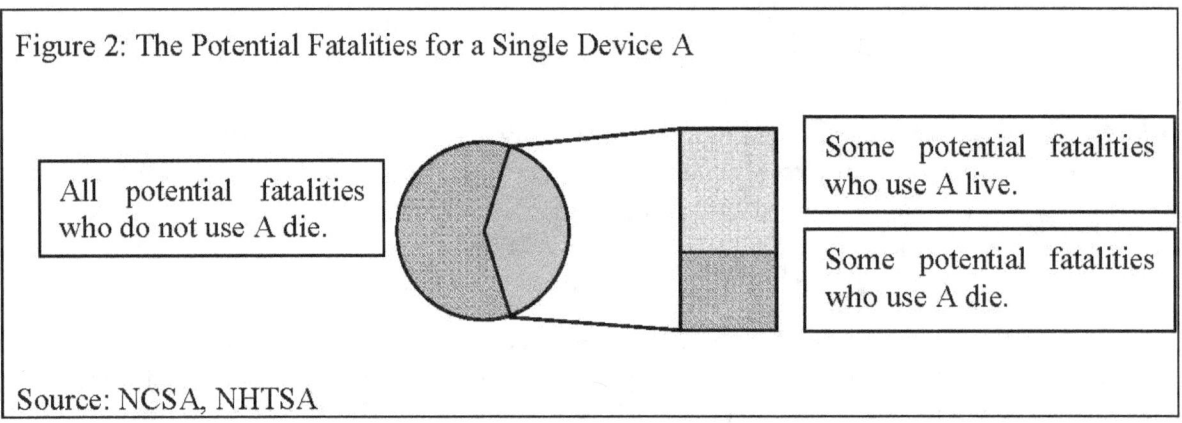

Figure 2: The Potential Fatalities for a Single Device A

All potential fatalities who do not use A die.

Some potential fatalities who use A live.

Some potential fatalities who use A die.

Source: NCSA, NHTSA

Note that the notion of potential fatality applies to the <u>person</u> experiencing the setting, not the particular instance of the setting that the person experiences. An instance of the setting can be potentially fatal to one person and survivable to another. For example, a particular frail elderly person might die in a crash in which a healthy young adult (in the same seating position) would have survived. However, the instance of the setting is sometimes called *potentially fatal* (e.g., potentially fatal crashes), with the understanding that this depends on the person experiencing the setting.

Note that the device is not rated on the entire population of people and instances of the setting, but only those in danger of dying. It would be disingenuous to rate helmets for motorcyclists in very minor crashes. Note also that the population on which the device is rated may include people who would die for reasons that have nothing to do with the setting or the device. For instance, the population against which motorcycle helmets are rated includes motorcyclists who died of impacts to the chest that occurred during the crashes.

Here, potential fatalities are a hypothetical population. However we also speak of a person who actually experiences the setting, with or without the single Device A, who would die without Device A as a potential fatality. Potential fatalities who live are generally not identifiable in particular instances of the setting. For example, we cannot say whether a helmeted motorcyclist who survives a particular crash would have died if the motorcyclist had not worn the helmet.

Every fatality that does not use Device A is a potential fatality. For example, for the device of motorcycle helmets in the setting of motorcycle crashes, all unhelmeted potential fatalities become fatalities when they experience their crashes.

4.1.3 Effectiveness Ratings

The *effectiveness* of the Device A is the proportion of the potential fatalities who would live if Device A had been used. For instance, the effectiveness of helmets is the percentage of motorcyclists who would survive crashes helmeted among those in crashes severe enough to kill them unhelmeted. Letting e denote the effectiveness of Device A, means that e×100 percent of the potential fatalities who use A live, while the others who use A die, and all who do not use A die. We assume in this report that devices increase the chance of survival, and so 0<e≤1. We note however that NHTSA does compute some negative effectiveness ratings, e.g., for children and air bags in Kahane (2004).

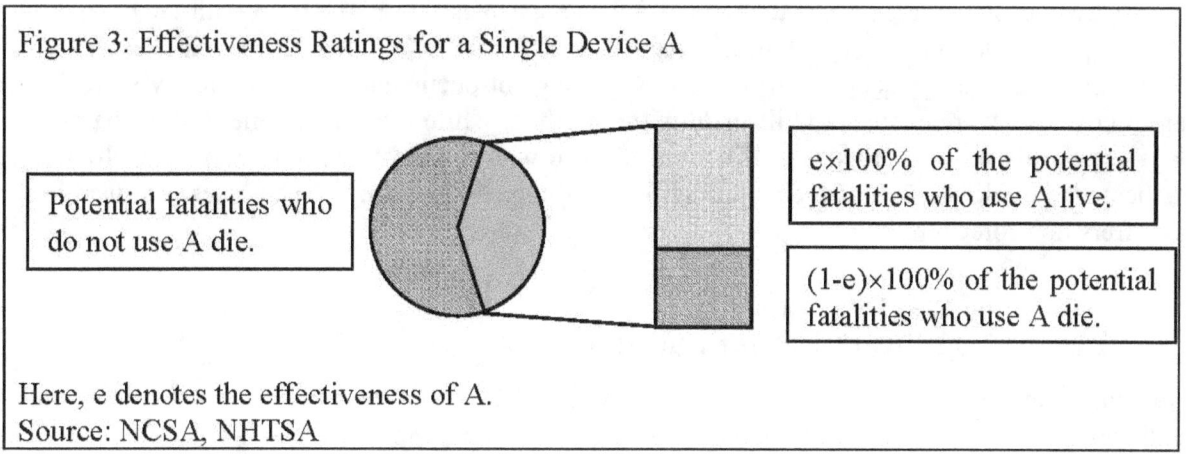

Figure 3: Effectiveness Ratings for a Single Device A

Potential fatalities who do not use A die.

e×100% of the potential fatalities who use A live.

(1-e)×100% of the potential fatalities who use A die.

Here, e denotes the effectiveness of A.
Source: NCSA, NHTSA

4.1.4 Estimated Effectiveness

Effectiveness is estimated from data. See Kahane (2000) for details on how this is done. For brevity, the estimated effectiveness of a device is frequently also called its *effectiveness*.

Because it is calculated from data, the estimated effectiveness represents the ability of A to protect the types of people and settings that occurred in the data set used. For example, the belt effectiveness ratings in Kahane (2000) represent the efficacy of belts in the types of crashes and people in crashes that occurred in the period 1986 – 1999, whose crash data was used to estimate this effectiveness. In particular, ratings may increase or decrease over time. If the nature of a crash were to suddenly change so that it was generally unsurvivable, the estimated belt effectiveness for that crash would become very small.

A device might be assessed different effectiveness ratings on different subpopulations. For instance, belts will be assessed for various vehicle types and seating positions, while NHTSA only rates air bag effectiveness for people over age 12.

4.1.5 Devices That Can Be Used Improperly

When a device can be misused, such as a belt buckled improperly, the device could be rated for its effectiveness as used or when used properly. Since effectiveness is estimated from data, devices are usually rated as used. This is the case for NHTSA's seat belt ratings, and so the estimates in Kahane (2000) reflect the extent to which people fail to buckle manual lap belts when they have automatic shoulder belts, for example, and the frequency with which motorists put shoulder belts behind their backs. If everyone used belts properly, the belt effectiveness ratings would be higher than those in Kahane (2000).

4.1.6 Passive Devices

When Device A is a passive device (i.e., it requires no action on the person's part to protect him/her) that may or may not be engaged in the setting (such as an air bag), use is frequently considered to constitute "presence," rather than "engagement" in the definition of effectiveness. However if the unengaged device is effectively useless, as is the case with air bags, the two ratings are the same. For instance, if bags were the sole safety devices in vehicles, the effectiveness of air bag <u>presence</u> would be the percentage of occupants who would survive with a bag in crashes severe enough to kill them without a bag, while the effectiveness of air bag <u>deployment</u> would be the percentage of occupants who would survive with a bag that deploys in crashes severe enough to kill them with a disabled bag. These are the same since a disabled bag offers no protection.

4.1.7 Estimating the Potential Fatalities

The number of potential fatalities using Device A in some time frame can be estimated from the fatalities using A in the time frame and the effectiveness rating. For example, if F people die in a year using a Device A that has the effectiveness e, then by the definition of effectiveness, $F/(1-e)$ people used A in an instance of the setting in which they would die unprotected in the same year. Note that the same formula $F/(1-e)$ can be applied using $e=0$ and taking F to be the unprotected fatalities to calculate the unprotected potential fatalities. So there are $F_1 + F_2/(1-e)$ potential fatalities, where F_1 fatalities did not use A and F_2 fatalities did.

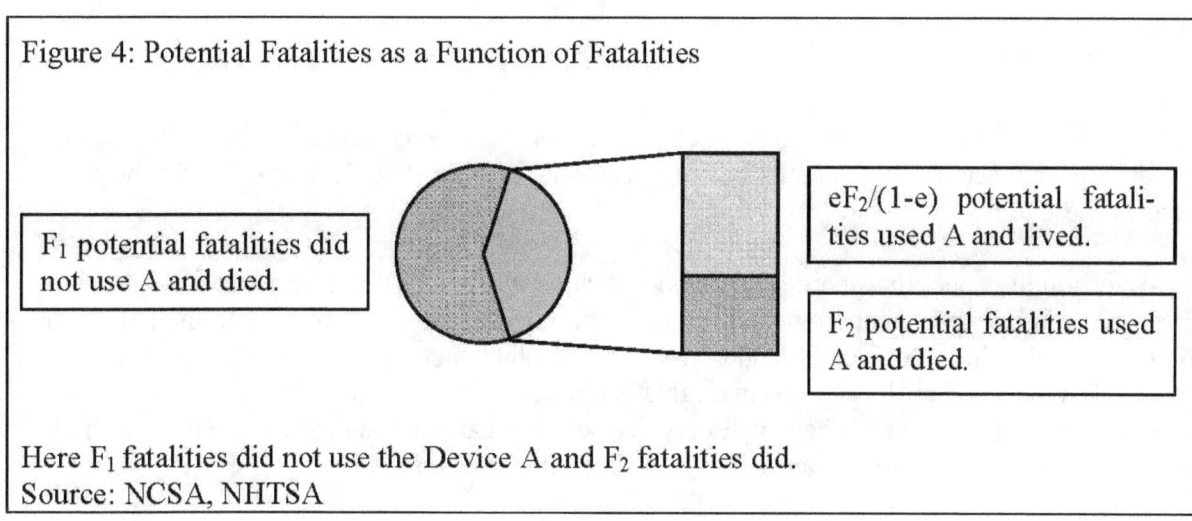

Figure 4: Potential Fatalities as a Function of Fatalities

F_1 potential fatalities did not use A and died.

$eF_2/(1-e)$ potential fatalities used A and lived.

F_2 potential fatalities used A and died.

Here F_1 fatalities did not use the Device A and F_2 fatalities did.
Source: NCSA, NHTSA

If one additionally knows the fraction, u, of the potential fatalities who used the devices, then one can alternatively calculate the number of potential fatalities in the time frame as follows. If F people died in the setting during the time frame (including those who used a device(s) and those who did not), then there were $F/(1-eu)$ potential fatalities, since eu of the potential fatalities lived and the rest died.

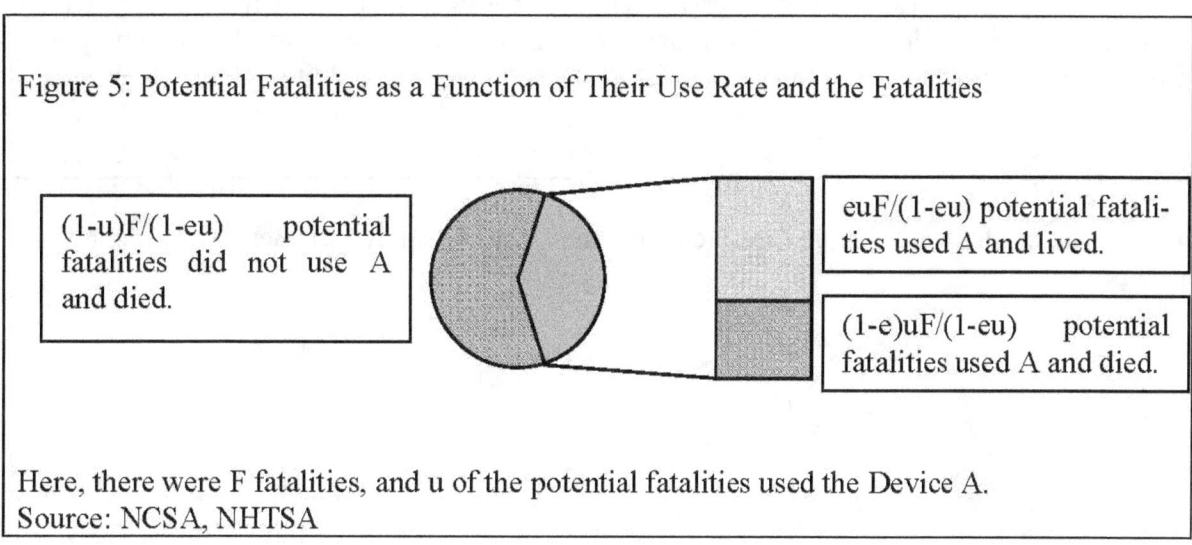

Figure 5: Potential Fatalities as a Function of Their Use Rate and the Fatalities

$(1-u)F/(1-eu)$ potential fatalities did not use A and died.

$euF/(1-eu)$ potential fatalities used A and lived.

$(1-e)uF/(1-eu)$ potential fatalities used A and died.

Here, there were F fatalities, and u of the potential fatalities used the Device A.
Source: NCSA, NHTSA

4.2   For Multiple Devices

The "sole-device" scenario is useful to illustrate the major concepts in benefits analysis. However, in actual life-threatening situations, such as vehicle crashes, usually a large number of factors affect survival, such as seat belts, air bags, front-disk brakes, weather conditions, and road conditions.

4.2.1 Devices and Settings

As explained in Section 4.1.1, the general calculation of lives saved involves specifying one or more <u>devices</u> that act in some potentially life-threatening <u>setting</u>. (See Section 4.1.1 for the definitions of these terms.)

The reader might expect that the devices would include <u>all</u> factors that affect survival. For instance, one might expect that front disk brakes, crumple zones, weather conditions, and road conditions would all be devices (in addition to several other factors) when computing the lives saved by seat belts and air bags. However, this is not generally the case. Our calculation of the lives saved by seat belts and air bags will only use seat belts and air bags as the devices, with all other factors, such as front disk brakes, being considered part of the setting.

The reader might also guess that the devices would consist precisely of the factors whose benefits one is interested in estimating. That will be the case for this report, in which we are interested in estimating the lives saved by seat belts and air bags, and will consider these two items as the devices.

For the remainder of our presentation of the concepts of effectiveness, potential fatalities, and lives saved (i.e., for the rest of this chapter, as well as Chapters 5 and 6), we will take it as given that the devices when calculating the lives saved by seat belts and air bags will be these two restraints, with all other factors, such as front disk brakes and weather conditions, considered part of the setting.

Figure 6: The Devices and Setting for Calculating the Lives Saved by Seat Belts and Frontal Air Bags

The Setting
Crashes of Passenger Vehicles

The Devices
Seat Belts and Air Bags

4.2.2 Potential Fatalities

For two Devices A and B protecting people in a common setting, the *potential fatalities* consist of people in the setting who would die if they experienced the setting having neither A nor B. For instance, viewing seat belts and air bags as a two-device scenario, people in crashes sufficiently severe that they would die unbelted with no air bag constitute the potential fatalities.

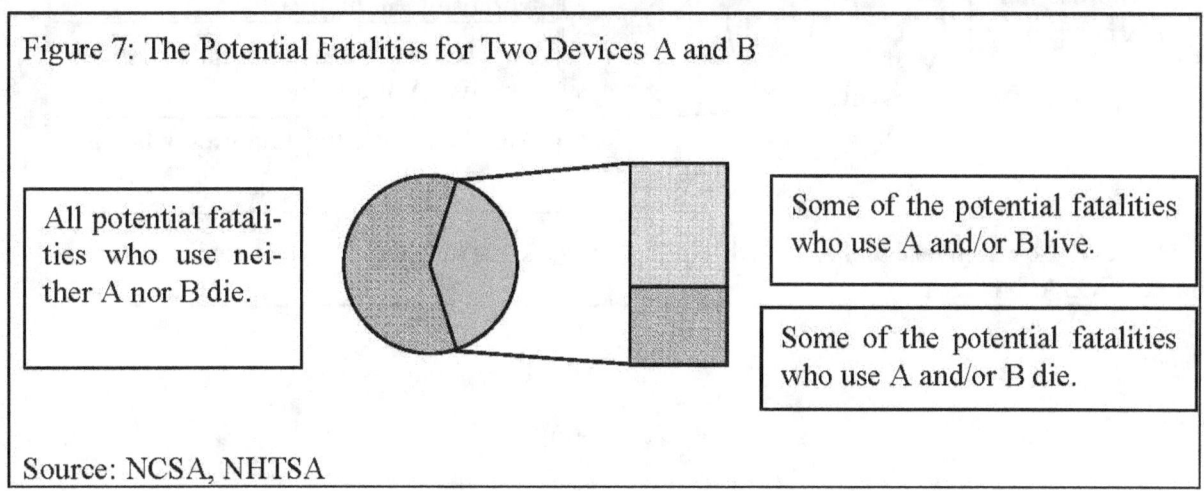

Figure 7: The Potential Fatalities for Two Devices A and B

All potential fatalities who use neither A nor B die.

Some of the potential fatalities who use A and/or B live.

Some of the potential fatalities who use A and/or B die.

Source: NCSA, NHTSA

As with a single device, the potential fatalities using neither A nor B are precisely the fatalities using neither device, and all fatalities were potential fatalities.

By rating against those who would die without either device, we can quantify how A and B affect each other. This will be captured in the joint and residual ratings in the next section.

4.2.3 Effectiveness Ratings

When two Devices A and B protect people in the same setting, there are a number of types of effectiveness ratings. The effectiveness e_A of A is the percent of potential fatalities who would live using A. That is, $e_A \times 100$ percent of people would live in the typical instance of the setting in which they would die without the use of either device. The joint effectiveness e_{AB} of A and B is percent of potential fatalities who would live if they used both A and B. The residual effectiveness $e_{B|A}$ of B is the percent of people who would live using B among those that would die using A alone.

Figure 8: Effectiveness for Two Devices A and B	
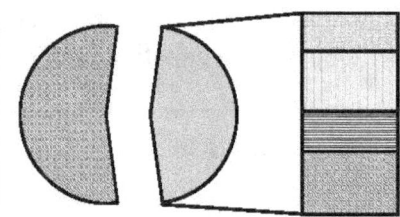 Potential fatalities who use neither A nor B die.	$e_A \times 100\%$ of the potential fatalities who use A alone live.
	$e_B \times 100\%$ of the potential fatalities who use B alone live.
	$e_{AB} \times 100\%$ of the potential fatalities who use A and B live.
	All other potential fatalities who use A and/or B die.

Here e_A is the effectiveness of A, e_B is that of B, and e_{AB} is that of A and B used in conjunction. Source: NCSA, NHTSA

Note that

$$e_{AB} = e_A + e_{B|A}(1 - e_A) \qquad\qquad (1)$$

That is, when both A and B are used, those not saved by A have an $e_{B|A} \times 100\%$ chance of being saved by B. That is, we remove B and see how many live (with A alone), and then add B back and see how many more will live. The relationship in (1) can also be written as

$1 - e_{AB} = (1 - e_A)(1 - e_{B|A})$, or
$e_{B|A} = \dfrac{e_{AB} - e_A}{1 - e_A}$.

Devices that Interact
In general, one device might enhance the efficacy of another (i.e., $e_{AB} > e_A + e_B$, i.e., A and B have a positive interaction), diminish it (i.e., $e_{AB} < e_A + e_B$, i.e., a negative interaction), or have no effect (i.e., $e_{AB} = e_A + e_B$, i.e., no interaction).

Devices frequently have the property that one (equivalently, each) device is no more effective on those who use the other device (than with those who do not use the other device). For example, air bags are less effective for belted motorists than for unbelted motorists. Note that in this case A and B have a nonpositive interaction, and a negative interaction if the devices have positive effectiveness. That is, if $e_{B|A} \leq e_B$ then $e_{AB} = e_A + e_{B|A}(1 - e_A) \leq e_A + e_B(1 - e_A) \leq e_A + e_B$, and the last inequality is strict if $e_A > 0$.

The Multiple Device Case as an Instance of the Single-Device Case
The effectiveness ratings (individual, joint, and residual) of multiple devices can be viewed as ratings of single devices on subpopulations of the single-device potential fatalities. If we view A and B as constituting a single device, then e_{AB} is the effectiveness of this device. The potential fatalities for this single device are precisely the potential fatalities for A and B as two devices.

For instance, the effectiveness of seat belts and air bags used jointly is the same, regardless of whether one considers these restraints to constitute two separate devices or one combined device.

The residual effectiveness of A is the effectiveness of the single Device A for persons using B. That is, $e_{A|B} \times 100\%$ of people using B who would die without A would survive with A. For example, the residual effectiveness of air bags is the effectiveness of air bags for belted motorists.

Similarly, the effectiveness of A as one of the two Devices A and B is the effectiveness of the single Device A for persons not using B, since e_A gives the proportion of survivors using A among persons protected by neither A nor B. For example, the effectiveness of air bags in the two-device scenario is the effectiveness of air bags as a single device for unbelted motorists.

4.2.4 Estimating the Potential Fatalities

The formulas from the single-device case are easy to extend to multiple devices. If F_A fatalities used A alone, F_B used B alone, F_{AB} used both, and F_0 used none, then the number of potential fatalities is

$$F_A/(1-e_A) + F_B/(1-e_B) + F_{AB}/(1-e_{AB}) + F_0 \qquad (2)$$

4.3   For Seat Belts and Frontal Air Bags

4.3.1 A Two-Device Scenario

Seat belts and frontal air bags are two of several technologies in vehicles (side air bags, crumple zones, reinforced passenger compartments, padded dashboards, etc.) designed to protect people in crashes. As we have indicated, we will consider these two restraints to constitute the devices when calculating the lives they save. All other possible contributing factors to crashes, such as front-disk brakes and driver distraction, will be considered part of the setting. Consequently, seat belts and air bags constitute a two-device scenario for us, and so the two-device theory from Section 4.2 applies.

Strictly speaking, crashes of vehicles that do not have air bags constitute a one-device setting. Note however, that this could be viewed as a two-device scenario, with the two devices being the seat belt and an "air bag," in which the "air bag" is ineffective (0% effective). This point of view will be convenient for our calculations.

4.3.2 Potential Fatalities

The potential fatalities in calculating the lives saved by seat belts and air bags consist of motorists over age 4 in vehicles on the road today who are in crashes sufficiently severe that they would die without either restraint.

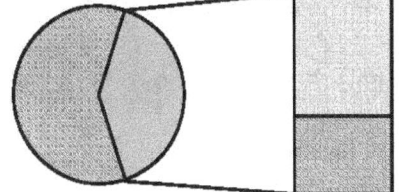

Figure 9: The Potential Fatalities for Seat Belts and Frontal Air Bags

Unbelted potential fatalities who do not have air bags die.		Some potential fatalities who use belts and/or have bags live.
		Some potential fatalities who use belts and/or have bags die.

Source: NCSA, NHTSA

Note that whether a person is a potential fatality depends on the person as well as the crash. A crash that is potentially fatal to an elderly person might be survivable to a younger person. Note also that the severity of a crash experience can depend on the seating position. For example, a driver-side crash might be fatal to a particular driver, who might have survived if the person had been in the right-front passenger's seat.

Regarding the age restriction that potential fatalities be older than 4, recall that NHTSA does estimate the numbers of children under age 5 who are saved by seat belts. This is done in a separate calculation documented in Starnes (2005).

4.3.3 Effectiveness Ratings

NHTSA's most recent ratings of seat belts and air bags are given in Tables 2 and 3, respectively. The two restraints are rated for different age groups, with belts rated for those 5 and older and bags for those over 12. Various types of belts (3-point, 2-point, etc.) are rated in each of the seating positions that have these belts. (. For example, if a vehicle has a front-center seat, it has a lap belt.) Frontal air bags are given an average rating for all types of frontal air bags and both applicable seating positions (driver and right-front passenger).

Table 2: Effectiveness Ratings for Seat Belts for Occupants 5 and Older						
Seating Position	**Front Left (Driver)**	**Front Right**	**Front Middle**	**Rear Outboard**	**Rear Middle**	**Other**
Passenger Cars						
2-Point	32%*	32%*	NA	NA	NA	NA
3-Point	48%	37%	NA	NA	NA	NA
Lap/Shoulder	48%[#]	37%[#]	NA	44%	NA	NA
Lap Belt	32%[#]	32%[#]	19%	32%	32%	32%
Unknown Type	32%[#]	32%[#]	19%[#]	32%[#]	32%[#]	32%[#]

18

		Light Trucks and Vans				
2-Point		NA	NA	NA	NA	NA
3-Point	61%	58%	NA	NA	NA	NA
Lap/Shoulder	NA	NA	NA	73	NA	NA
Lap Belt	NA		32%	63%	63%	63%
Unknown Type	61%[#]	58%[#]	32%[#]	63%[#]	63%[#]	63%[#]

[*] These ratings were used instead of the published rating on the advice of the author of Kahane (2000).
[#] The belts in these cells have not been rated. These estimates are used on the advice of the author of Kahane (2000).
There are no belts of the indicated type in cells labeled "NA."
Sources: Kahane, 2000, and Morgan, 1999.

Note the ratings given in Table 2 for belt types that have not been specifically rated. When not rated, lap belts are given the same effectiveness as two-point belts and lap/shoulder belts that of 3-point belts. Unknown belt types are given the effectiveness of lap belts.

We made the following effectiveness assignments for FARS cases with unknown seating position. Occupants known to be in the front seat, but whose specific seating position (driver, right-front passenger, or center-front) is unknown, were assigned the effectiveness of the right-front passenger with unknown belt type, and similarly for rear-seat occupants. Occupants whose seating position was completely unknown were assigned the rear-seat effectiveness. It is important to note that these are conservative assignments.

Recall that the joint belt-bag ratings can be obtained from Equation (1). For instance, 3-point belts are 48 percent effective for occupants over 4 years old in the driver's seat of passenger cars. They are 54 percent effective in conjunction with air bags, since by Equation (1), $0.48 + 0.11 \times (1-0.48) = 0.5372$, or 0.54 when rounded. Table 4 gives the joint effectiveness ratings of belts and bags for occupants over 12.

Table 3: Effectiveness Ratings of Frontal Air Bags for Occupants Over 12 Years Old	
Effectiveness of frontal air bags	14%
Residual effectiveness of frontal air bags	11%
Source: Fifth/Sixth Report to Congress, 2001	

Table 4: Effectiveness Ratings for Seat Belts in Conjunction With Frontal Air Bags for Occupants Over 12 Years Old						
Seating Position	Front Left (Driver)	Front Right	Front Middle	Rear Outboard	Rear Middle	Other
Passenger Cars						
2-Point	39%	39%	NA	NA	NA	NA
3-Point	54%	44%	NA	NA	NA	NA
Lap/Shoulder	54%	44%	NA	44	NA	NA
Lap Belt	39%	39%	19%	32%	32%	32%
Unknown Type	39%	39%	19%[#]	32%[#]	32%[#]	32%[#]
Light Trucks and Vans						
2-Point	NA	NA	NA	NA	NA	NA
3-Point	65%	63%	NA	NA	NA	NA
Lap/Shoulder	NA	NA	NA	73	NA	NA
Lap Belt	NA	NA	32%	63%	63%	63%
Unknown Type	65%	63%	32%[#]	63%[#]	63%[#]	63%[#]
There are no belts of the indicated type in cells labeled "NA." Sources: Kahane (2000) and Morgan (1999)						

Recall that effectiveness ratings indicate the proportion of potential fatalities who will live if they use the devices. Figures 10 and 11 depict the ratings for passengers with 3-point belts in the right-front seating position in passenger cars. Note the different treatment of motorists over 12 years old, in Figure 10, versus those between 5 and 12, in Figure 11. This difference is based on the fact that air bags are considered effectively neutral for children 5 to 12. Note that the 44 percent effectiveness shown below in Figure 10 was calculated using Equation (1), where $e_{AB} = e_A + e_{B|A} (1 - e_A)$, and thus $0.37 + 0.11 \times (1-0.37) = 0.4393$, or 44 percent when rounded; by comparison, the 37 percent effectiveness shown below in Figure 11 was calculated using Equation (1), where $e_{AB} = e_A + e_{B|A} (1 - e_A)$, and thus $0.37 + 0.00 \times (1-0.37) = 0.37$, or 37 percent.

Figure 10: Effectiveness of 3-Point Belts and Frontal Air Bags for Passengers Over 12 in the Right-Front Seats of Passenger Cars

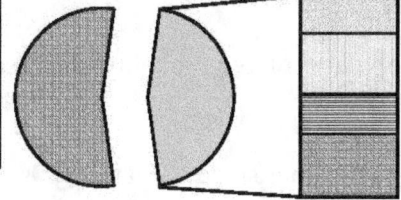

Unbelted poten-tial fatalities who do not have air bags die.		37% of the belted potential fatalities who do not have air bags live.
		14% of the unbelted potential fatali-ties who have air bags live.
		44% of the belted potential fatalities who have air bags live.
		All other potential fatalities who use belts and/or have air bags die.

Source: NCSA, NHTSA, and Kahane (2000).

Figure 11: Effectiveness of 3-Point Belts and Frontal Air Bags for Passengers 5 to 12 in the Right-Front Seats of Passenger Cars

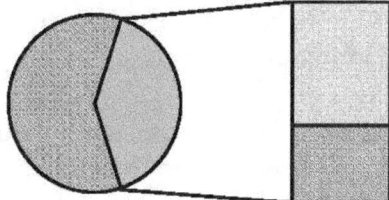

| Unbelted poten-tial fatalities in this age group die. | | 37% of the belted potential fatalities live. |
| | | 63% of the belted potential fatalities die. |

Source: NCSA, NHTSA, and Kahane (2000).

Air bags are also considered effectively neutral (e=0.00) for occupants of the center seating posi-tion in the front seat. This is because the extent to which frontal air bags protect the front center seating position is unknown and the agency has not estimated air bag effectiveness for this posi-tion. Therefore, for the purpose of the estimates in this report, we assume they offer no protec-tion.

The ratings in Kahane (2000) and Morgan (1999) were computed using data from 1986 – 1999 and 1988 – 1997 respectively, and those in (Fifth/Sixth Report to Congress, (2001) used 1986 – 2000. Consequently NHTSA's current seat belt and air bag ratings reflect the crash and motorist characteristics from the late 1980s and the 1990s. For instance, seat belts greatly improve the chance for survival in rollover crashes. The higher belt effectiveness ratings in LTVs compared to cars probably reflects the greater frequency of rollovers (on the basis of vehicle miles traveled, VMT) among SUVs compared to cars. Since SUVs and rollover crashes have become even more prevalent since belts were rated, the belt ratings in LTVs might be too low for today's crashes.

Vehicle characteristics have also changed since the late 1980s and the 1990s. Today's restraints have improved (e.g., because of belt pretensioners and advanced -- "smart" -- air bag systems) and so the current belt and bag effectiveness ratings might underestimate the effectiveness in later model vehicles.

Similarly, the demographics of the motorist population have probably changed since the 1990s. The effect of this on the effectiveness ratings is unclear.

Alcohol involvement in crashes has decreased somewhat since the late 1980s, and the typical alcohol-involved crash is more severe than the typical sober-driver crash. Crashes involving alcohol are 50 percent more likely to result in an injury or fatality than non-alcohol crashes. (NCSA, *2001 Annual Assessment of Motor Vehicle Crashes*, 2002) In 2002, 6 percent of crashes involved alcohol, but these crashes accounted for 41 percent of the traffic fatalities (NCSA, *Traffic Safety Facts 2002 – Alcohol*, undated). The impact of this on the effectiveness ratings is unclear.

Because the seat belt ratings were estimated at a time when there were virtually no air bags, the belt ratings in Kahane (2000) and Morgan (1999) were effectively derived using a one-device scenario. However, as we noted earlier, since the individual effectiveness of a device in a two-device scenario agrees with its effectiveness as a single device for people who do not have the other device, the belt ratings in Kahane (2000) and Morgan (1999) can be viewed as the individual effectiveness ratings of seat belts in the two-device belt-and-bag scenario. The air bag ratings in the Report to Congress (2001) were obtained as two-device ratings.

Likewise the seat belt and air bag ratings were rated in vehicles largely without side air bags. Consequently we do not have residual ratings for belts and bags conditioned on the presence of side air bags. That is, we do not have numbers to show how effective belts and bags are in vehicles that have side air bags.

Recall that since disabled air bags offer no protection, one can view the effectiveness ratings of air bags as rating either the presence or the deployment of a bag. That is, air bags are 14 percent effective when they deploy and a bag's presence is just as effective (although this may initially seem odd). Also recall that the 14 percent effectiveness rating of air bags is the effectiveness of air bags for unbelted occupants, while bags are 11 percent effective for belted occupants. Note that these are average ratings that apply to all vehicle types, to both applicable seating positions (i.e., both front outboard positions), and all belt types.

5.   Example 1 and Notation

The previous chapter introduced key preliminary notions necessary for a rigorous computation of saved and savable lives. We are now at a point where we can begin making some computations and developing formulas. Consequently it is convenient at this point to pause to establish some notation and provide an example that we will use to illustrate the computations. In the next chapter we resume the presentation of how to compute saved and savable lives.

5.1 Example 1

Table 5 presents some fatality counts from the 2002 FARS Annual Report File (ARF), as well as effectiveness ratings of the restraints used by motorists (over age 4) in the right-front seat of passenger cars with 3-point belts in 2002. This example uses fatality data from the FARS 2002 ARF; how-

| Table 5: Fatalities and Effectiveness Ratings in 2002 Among Right-Front Passengers in Cars With 3-Point Belts (Example 1) |||||
Age 5-12?	Belt Used?	Air Bag in Seating Position?	Effectiveness of Restraint Used	Fatalities
Yes	Yes	NA	37%	57
Yes	No	NA	0%	26
No	Yes	Yes	43.93%	1,110
No	Yes	No	37%	669
No	No	Yes	14%	882
No	No	No	0%	837
Total*				3,581
***Items might not sum to totals due to rounding.** **Source: NCSA, NHTSA, FARS, 2002**				

ever, please note that these effectiveness ratings were used by NHTSA in 2002, and each effectiveness rating is still being used by NHTSA at the time of this report. See Kahane (2000) and Morgan (1999) for more details.

This example includes a specified seat position, vehicle type, restraint use, and year. Our calculations apply only to passengers in this right-front seat position, vehicle type (passenger car), and restraint use (3-point belts), as different effectiveness ratings apply to different combinations of these variables.

We will refer to this as Example 1 throughout this report, and will use it to illustrate the calculations throughout this report. Note that all belted children under 13 are given the same effectiveness ratings, whether or not they were seated in front of air bags, since the effectiveness for air-bags is 0.00 for children under 13. Note also that the 43.93 percent effectiveness of the combined belt-and-bag system is related to the 37 percent effectiveness of the belt and the 11 percent residual effectiveness of the bag according to the formula $e_{AB} = e_A + e_{B|A} (1 - e_A)$ from the previous section.

In the several instances in which we use Example 1 in this report, items may not sum to totals due to rounding.

5.2 Notation

In this section we present notation that will be used throughout this report.

We will derive our formulas for saved and savable lives first in the context of a general safety device A, then for two devices, A and B (which will indicate how three or more devices would be handled), and finally for belts and bags.

Notation Used for a Single Device A

The notation in Table 6 will be used in the one-device setting.

Table 6: Notation for a Single Safety Device A	
Notation	Definition
e	the effectiveness of A against fatality
F	the fatalities occurring in some time period that used A
P	the potential fatalities (with respect to A) that occurred in the same time period
x_{hypoth}	a hypothesized use rate for A
x_{actual}	the actual use rate for A
u_{hypoth}	the use rate (for A) among the potential fatalities that occur when the use rate in the general population is x_{hypoth}
u_{actual}	the actual use rate for A among potential fatalities

Notation for Two Devices, A and B

The following notation will be used in the context of two devices, A and B. Note that the use rates in this table pertain only to the use of A, not of B.

Table 7: Notation for Two Safety Devices, A and B	
Notation	Definition
e_A (or e_B, e_{AB})	the effectiveness of A (or B, the combination of A and B) against fatality
$e_{A|B}$ (or $e_{B|A}$)	the residual effectiveness of A (or B)
F_A (or F_B, F_{AB}, F_0)	the fatalities occurring in some time period that used A alone (or B alone, A and B, neither device)
P_A (or P_B, P_{AB}, P_0)	the potential fatalities (with respect to A and B) that occurred in the same period and used A alone (or B alone, A and B, neither device)
x_{hypoth}	a hypothesized use rate
x_{actual}	the actual use rate
u_{hypoth}	the use rate among the potential fatalities that occur when the use rate in the general population is x_{hypoth}
u_{actual}	the actual use rate among potential fatalities

Notation for Seat Belts and Frontal Air Bags

Table 8 presents the notation used for belts and frontal air bags. Recall that "restraint configurations" were defined in Chapter 3.

Table 8: Notation for Seat Belts and Frontal Air Bags	
Notation	**Definition**
e(bag)	The effectiveness of air bags, i.e., 14 percent
e(bag | belt)	the residual effectiveness of air bags, i.e., 11 percent
R	the set of all restraint configurations
In the remaining definitions, i denotes a restraint configuration.	
F_i	the fatalities with restraint configuration i that occurred in some time period
belt(i)	1 if a belt is used in i, and 0 otherwise
bag(i)	1 if a bag is present and the occupant is over 12 in i, 0 otherwise
e_i(belt)	the effectiveness of the belt in i
e_i(system)	e_i(belt) if bag(i)=0, otherwise the effectiveness of the belt-bag system in i
e_i(used)	the effectiveness of the restraint (belt, bag, or belt-bag) used in i, 0 if unrestrained, with bags treated as 0 percent effective for children under 13
e_i(belt | bag)	the residual effectiveness of the belt in i, defined only when bag(i)=1

For instance, with this notation we have

$$e_i \text{(used)} = \begin{cases} e_i \text{(system)} & \text{if belt(i)} = 1 \text{ and bag(i)} = 1 \\ e_i \text{(belt)} & \text{if belt(i)} = 1 \text{ and bag(i)} = 0 \\ e\text{(bag)} & \text{if belt(i)} = 0 \text{ and bag(i)} = 1 \\ 0 & \text{if belt(i)} = 0 \text{ and bag(i)} = 0 \end{cases}$$

Note that because we defined bag(i) to be 0 for children under 13, e_i(used) reflects the belt alone for this age group, even if the seating position has an air bag. That is, e_i(used) is the effectiveness of the restraint(s) used, treating air bags as effectively neutral for children under 13. Similarly e_i(system) is the effectiveness of the system the occupant could have used, had they chosen to use all available restraint(s), treating air bags as 0 percent effective for young children.

The fifth data row in Table 5 describes the restraint configuration i=(passenger car, right-front passenger, 3-point, no, yes, 13 or older), and

belt(i)=0,
bag(i)=1,
F_i = 882
e_i(belt) = 37%
e_i(system) = 43.93%
e_i(used) = e(bag) = 14%, and
e(bag \| belt) = 11%.

Further, e_i(belt | bag) is defined since bag(i)=1, and has the value (e_i(system)-e(bag))/(1-e(bag)) = 34.8%.

When bag(i)=0, the occupant is either under 13 or is in a vehicle that does not have an air bag in that seating position. Such an occupant actually experiences a one-device scenario.

5.2.1 Estimating the Potential Fatalities for Seat Belts and Frontal Air Bags

For each belt type A for which we have a rating, we could apply the formula below with this belt type A and taking B to be a frontal air bag to compute the potential fatalities who had this belt type. We could then sum over the belt types to obtain the total potential fatalities. This is the approach we will take.

Formula

Using the notation from Section 5.2, the formula for the potential fatalities from (2) for belts and bags is then $\sum_{i \in R} \dfrac{F_i}{1 - e_i(\text{used})}$. Note that this formula treats air bags as effectively neutral for children under 13.

Example 1

This formula is illustrated in Table 9 (which uses data that originated in Table 5), finding that 5,021 right-front passengers in passenger cars equipped with 3-point belts were potential fatalities in 2002. Note that this calculation uses the joint effectiveness for occupants that both were belted and had an air bag.

Age 5-12?	Belt Used?	Air Bag in Seating Position?	Effectiveness of Restraint Used	Fatalities	Potential Fatalities
Yes	Yes	NA	37%	57	91
Yes	No	NA	0%	26	26
No	Yes	Yes	43.93%	1,110	1,980
No	Yes	No	37%	669	1,062
No	No	Yes	14%	882	1,026
No	No	No	0%	837	837
Totals*				3,581	5,021

Table 9: Potential Fatalities Among Right-Front Passengers in Cars With 3-Point Belts in 2002

*Items may not sum to totals, due to rounding.
Source: NCSA, NHTSA, FARS, 2002

6.   Lives Saved

In this section, we define what we mean when we say that a person was "saved" by a device, and derive how the number saved is estimated. This is complex for two devices if one wants to say which device saved a person, and not just that s/he was saved by at least one of the two. We first discuss the simpler one-device scenario, then the two-device scenario, and finally the application of the two-device scenario to seat belts and frontal air bags.

6.1   For a Sole Device

For a sole Device A, we say that a person in the setting was *saved by A* if s/he used A, survived, and would have died had s/he not used A. That is, the people saved by A are the potential fatalities who survive. Using the notation of Section 5.2, the number of lives saved is eP, or eF/(1-e). Equivalently, one can think of the number of saved lives as the number of potential fatalities F/(1-e) minus the number of actual fatalities F.

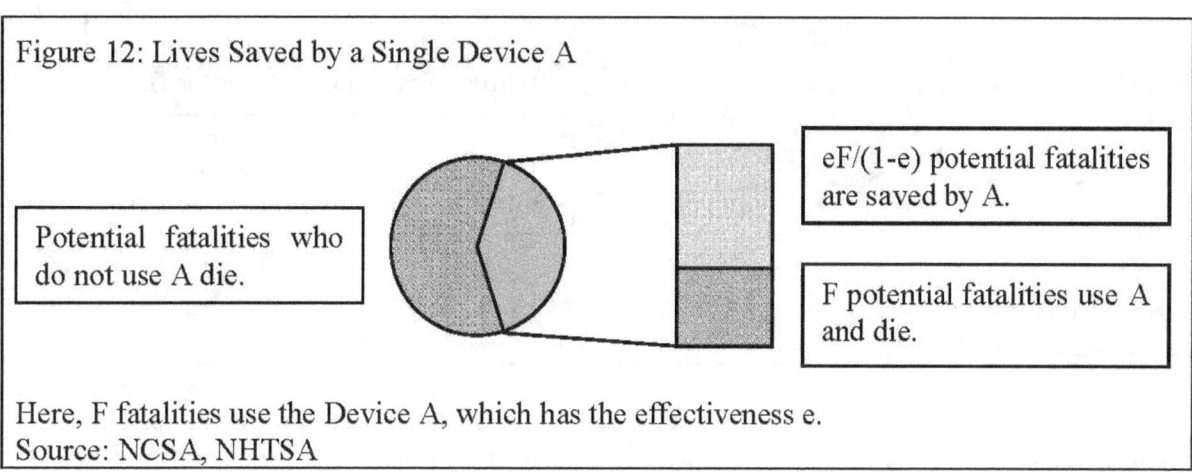

Figure 12: Lives Saved by a Single Device A

Potential fatalities who do not use A die.

eF/(1-e) potential fatalities are saved by A.

F potential fatalities use A and die.

Here, F fatalities use the Device A, which has the effectiveness e.
Source: NCSA, NHTSA

Throughout the estimation of the number of lives saved, it is important to remember that not everyone who lives using the devices are saved by them, only those who were in danger of dying. We cannot identify which survivors owe their lives to the devices.

Note that since the function eF/(1-e) is an increasing function of e on the domain 0<e<1, underestimated effective ratings result in underestimated lives saved.

6.2 For Multiple Devices

6.2.1 The Total Lives Saved

For Devices A and B, we say that a person in the setting was *saved by A and/or B* if the person used at least one of A and B, survived, and would have died had the person used neither A nor B. Using the notation of Section 5.2, the number of lives saved by A and/or B is

$$e_A P_A + e_B P_B + e_{AB} P_{AB}$$

or, in terms of fatalities,

$$\frac{e_A F_A}{1-e_A} + \frac{e_B F_B}{1-e_B} + \frac{e_{AB} F_{AB}}{1-e_{AB}} \; .$$

Figure 13: Lives Saved by Two Devices

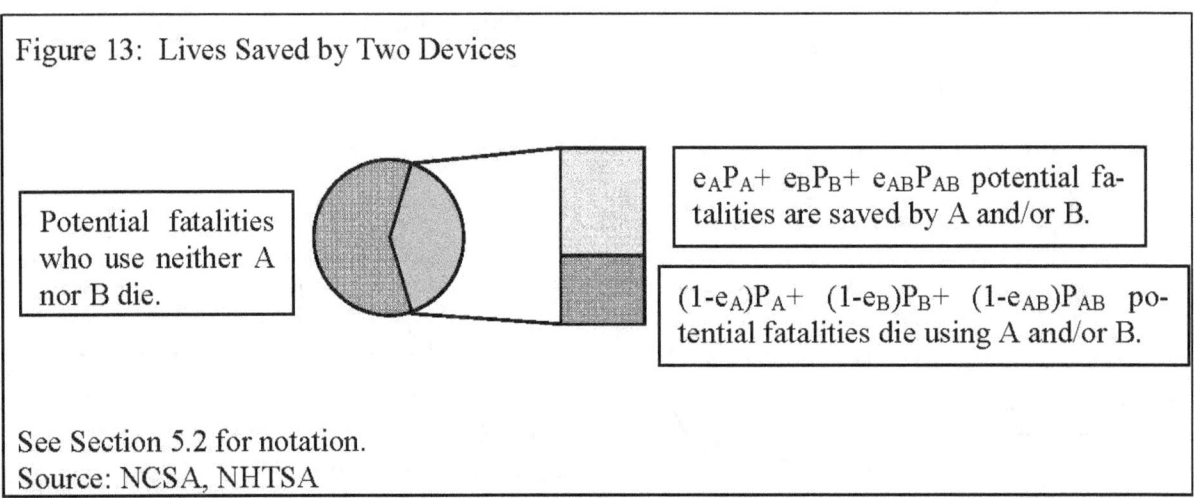

Potential fatalities who use neither A nor B die.

$e_A P_A + e_B P_B + e_{AB} P_{AB}$ potential fatalities are saved by A and/or B.

$(1-e_A)P_A + (1-e_B)P_B + (1-e_{AB})P_{AB}$ potential fatalities die using A and/or B.

See Section 5.2 for notation.
Source: NCSA, NHTSA

When factors that contribute to survival are considered part of the setting (as, for example, crumple zones are when calculating the lives saved by seat belts and air bags), they are not credited with saving lives. For instance, if B is considered part of the setting, a person who requires both A and B to live would be a potential fatality from the point of view of considering B as part of the setting. If this person used A and B, then the one-device viewpoint would credit the person's life to A. The -device viewpoint would credit that life to A and/or B.

Obviously a person who only used one of the two Devices A and B, must have been saved by it. However, saying who was saved by what among people who use A and B is complicated. Survival might depend on A but not B, on B but not A, on both, or either of the devices might be sufficient for survival.

Figure 14, below, shows the categories that exist for occupants who were potential fatalities but survived the crash. These categories are based on whether Device A and/or Device B were used, and whether Device A and/or Device B were needed for the occupant to survive. Figure 13 is only a qualitative summary, and does not provide any formulas for estimating lives saved.

Section 6.3 applies the two-device viewpoint toward estimating lives saved by seat belts and frontal air bags.

Figure 14: The Surviving Potential Fatalities and What Saved Them

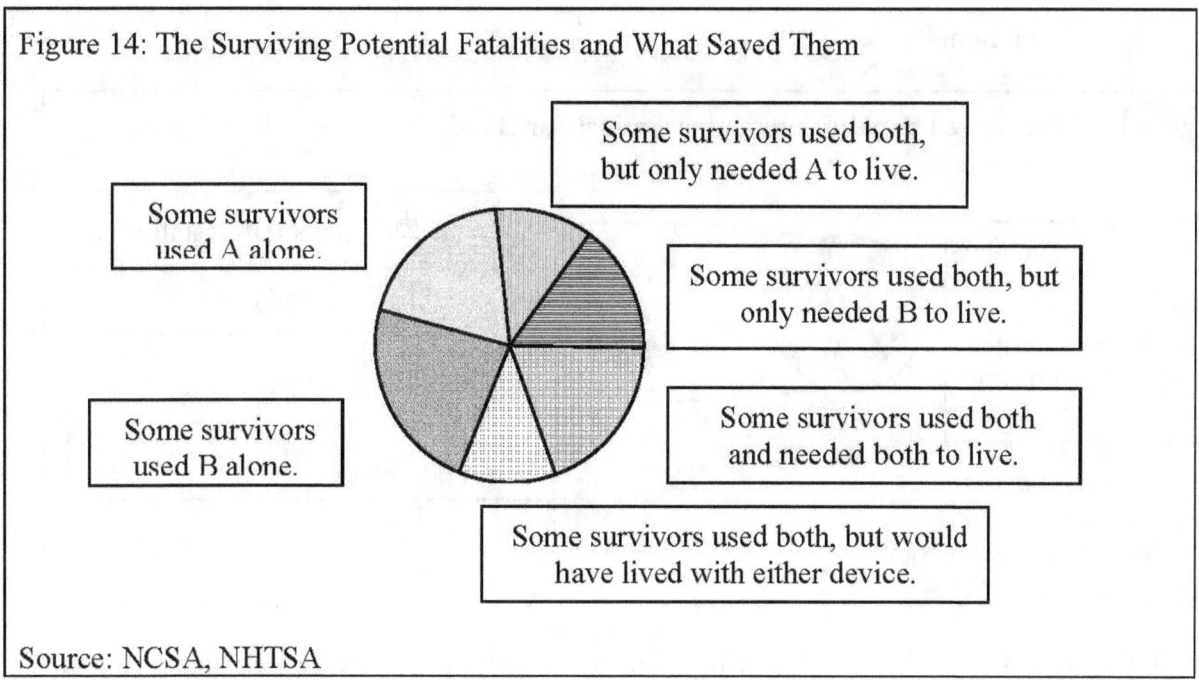

Some survivors used both, but only needed A to live.

Some survivors used A alone.

Some survivors used both, but only needed B to live.

Some survivors used B alone.

Some survivors used both and needed both to live.

Some survivors used both, but would have lived with either device.

Source: NCSA, NHTSA

6.3   For Seat Belts and Frontal Air Bags

Following the two-device scenario, a vehicle occupant in a crash is said to be *saved by a seat belt and/or frontal air bag* if s/he used at least one of these restraints, survived, and would have died had s/he used neither restraint.

Recall that we consider all factors contributing to crash survival other than seat belts and air bags to be part of the setting of vehicle crashes. Consequently, when a motorist's survival depends at least in part on a seat belt or an air bag, the motorist's survival is attributed to the belt or bag, regardless of whether other factors were also necessary for survival. See Appendix, Section 1, and particularly Appendix Section 1.4, for additional information

6.3.1 The Total Lives Saved

As with the calculation of potential fatalities, we apply the two-device formula to each belt type (Device A) and air bag (Device B). This gives the following.

Formula
The lives savable by a seat belt, an air bag, or both, are:

lives saved by belts and/or bags $= \sum_{i \in R} \dfrac{e_i \, (\text{used}) \, F_i}{1 - e_i \, (\text{used})}$

using the notation from Section 5.2.

Figure 15: Lives Saved by Seat Belts and/or Frontal Air Bags

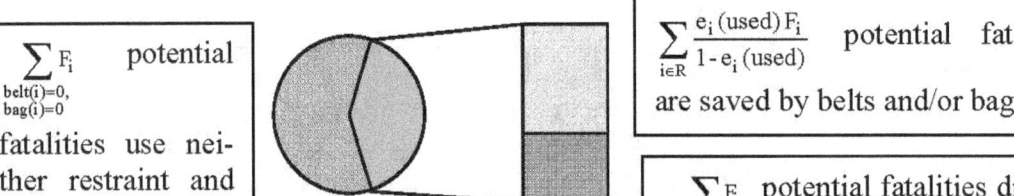

$\sum_{\substack{\text{belt(i)=0,} \\ \text{bag(i)=0}}} F_i$ potential fatalities use neither restraint and die.

$\sum_{i \in R} \dfrac{e_i \, (\text{used}) \, F_i}{1 - e_i \, (\text{used})}$ potential fatalities are saved by belts and/or bags.

$\sum_{\substack{\text{belt(i)=1 or} \\ \text{bag(i)=1}}} F_i$ potential fatalities die using belts and/or bags.

See Section 5.2 for notation.
Source: NCSA, NHTSA

Recall that from Section 4.3.3, that the effectiveness ratings we are using are reasonably expected to underestimate effectiveness in today's crashes. This would result in lives saved being underestimated as well.

Example 1
Applying the above formula to each line of Example 1 gives that 1,440 right-front passengers in cars equipped with 3-point belts were saved in 2002. Note that we are using the joint effectiveness for occupants who both were belted and had air bags.

Table 10: Lives Saved Among Right-Front Passengers in Cars With 3-Point Belts in 2002

Age 5-12?	Belt Used?	Air Bag in Seating Position?	Effectiveness of Restraint Used	Fatalities	Potential Fatalities	Lives Saved
Yes	Yes	NA	37%	57	91	34
Yes	No	NA	0%	26	26	0
No	Yes	Yes	43.93%	1,110	1,980	870
No	Yes	No	37%	669	1,062	393
No	No	Yes	14%	882	1,026	144
No	No	No	0%	837	837	0
Totals*				3,581	5,021	1,440

*Items may not sum to totals, due to rounding.
Source: NCSA, NHTSA, FARS, 2002

Note that the number of lives saved can also be calculated by subtracting the number of fatalities from the number of potential fatalities.

The Calculation Nationwide
Applying the same calculation nationwide yields that 16,441 occupants were saved by belts or bags in 2002.

State Calculations

The same formula can be used to estimate the lives saved in each State. However these numbers will not necessarily sum to the lives saved nationwide. Although the formula for saved lives is additive, there are distributions of unknowns, such as unknown restraint use, at the State and national levels, that prevent the applications of the formula from being additive.

Consequently we adjust the State numbers to total to the lives saved nationwide. We benchmark to the national total because the effectiveness estimates were derived from national crash data rather than crash data from individual States.

6.3.2 Attribution

In addition to estimating the lives saved by seat belts and air bags combined, NHTSA wishes to parcel this quantity into the number saved by seat belts and the number saved by air bags. These numbers are used in cost-benefit analyses of proposed seat belt or air bag regulations. They also appear in publications, pamphlets, and speeches concerning occupant protection.

Attributing survival to two safety Devices A and B is complex. Survival might depend on A but not B, on B but not A, on both, or either of the devices might be sufficient for survival. NHTSA partitions the lives saved into two quantities: those saved by seat belts and those saved by air bags.

Among occupants who are belted and positioned in front of air bags, some required both devices for survival and others needed only one device for survival. It is not possible to know which device saved the lives of each individual occupant, as no database can provide the plethora of information that would be need to produced exact lives saved counts. This leads NHTSA to produce lives saved estimates, rather than exact counts. For those occupants who are belted and positioned in front of air bags, the effectiveness estimates will be used to proportionally weight the lives saved estimates between seat belts and air bags, effectively attributing them to one device or the other.

Because air bags are effectively rated as neutral for children under 13, all surviving potential fatalities under 13 are necessarily belted and are therefore attributed to this restraint. The more complex case arises for potential fatalities over age 12 and is depicted in the next figure.

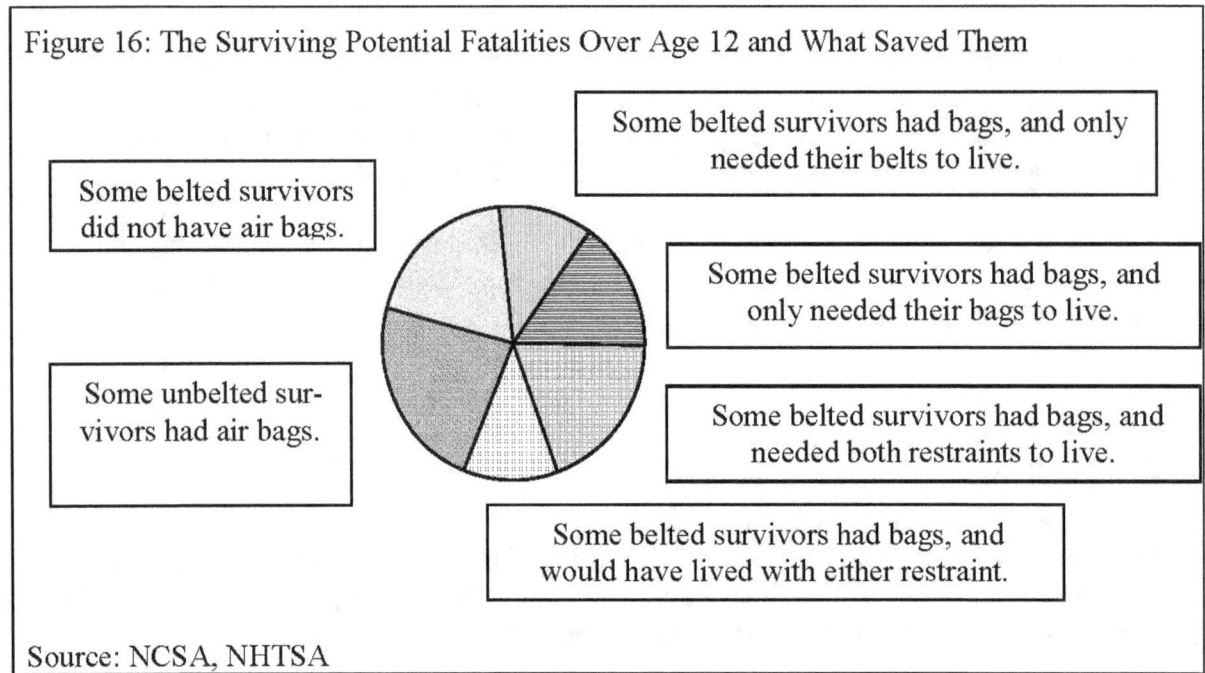

Figure 16: The Surviving Potential Fatalities Over Age 12 and What Saved Them

Some belted survivors did not have air bags.

Some belted survivors had bags, and only needed their belts to live.

Some belted survivors had bags, and only needed their bags to live.

Some unbelted survivors had air bags.

Some belted survivors had bags, and needed both restraints to live.

Some belted survivors had bags, and would have lived with either restraint.

Source: NCSA, NHTSA

Formulas

We apply the lives saved formulas to the two-device seat belt and air bag scenario in each restraint configuration. This method attributes

$$\sum_{\substack{belt(i)=1, \\ bag(i)=0}} \frac{e_i\,(belt)\,F_i}{1 - e_i\,(belt)} + \sum_{\substack{belt(i)=1, \\ bag(i)=1}} \frac{e_i\,(belt)\,F_i}{1 - e_i\,(system)}$$

of the lives saved to seat belts. The first summation above applies to belted occupants without an air bag (listed at belt = 1, bag = 0). The second summation above applies to belted occupants with an air bag (listed as belt=1, bag=1). See Section 5.2 for notation. Note that the effectiveness in the numerator of the second summation is the effectiveness of the seat belt, and the effectiveness in the denominator of the second summation is the joint effectiveness of the seat belt and air bag, referred to as "system."

Note that the surviving children age 5 to 12 all appear in the first sum, since we defined bag(i) to be 0 for this age group, regardless of whether there was an air bag in that seating position. The lives saved among these children are attributed to seat belts, as they should be under NHTSA's current air bag ratings.

The above formula attributes to belts all belted surviving potential fatalities for whom the belt was sufficient for survival. All other surviving potential fatalities (namely, those potential fatalities for whom the bags were necessary for survival) would be attributed to air bags. For example, a belted surviving potential fatality with an air bag who needed both restraints to live would be saved by the air bag.

32

This attribution is the method used in conventional benefits analysis. Conventionally, the safety device that is instituted first (in this case, belts) is credited with saving those people who would have lived with this device alone, and subsequent safety devices (in this case, bags) are attributed only the residual benefits. In addition, this method, called the "Reverse Chronological" method in Kahane (2004), was required by OMB Circular A-4 (Office of Management and Budget, 2003) for the cost-benefit analysis for seat belts calculated for Federal Motor Vehicle Safety Standard (FMVSS) No. 208. We also note that air bags are intended to supplement seat belts, in that NHTSA recommends that motorists buckle up to properly position themselves for the air bags. In this sense, air bags should only be accorded the residual benefits.

The formula for the lives saved by seat belts can be simplified to

$$\sum_{belt(i)=1} \frac{e_i \, (belt) \, F_i}{1 - e_i \, (used)} \, .$$

Example 1

We illustrate the limits and choices on Example 1, right-front passengers in cars with 3-point belts in 2002. See Table 10 for reference.

Obviously the 393 people over 12 in Table 10 who lived and did not have an air bag must have been saved by the seat belts. (Keep in mind that this is an estimated number of survivors. We do not know who the survivors are, and few if any would be in FARS even if we knew who they were.) Similarly, the 34 children under 13 who survived were also saved by the seat belts, since according to NHTSA's current estimates, air bags are effectively neutral for this age range. The 144 people who were unbelted and saved were saved by the air bags. (Presumably the air bags deployed since we expect undeployed air bags to offer no protection.) The only issue is how to attribute the 870 who were saved while being protected by both restraints.

Since their belts are 37 percent effective, belts were sufficient for the survival of 732 of the 1,980 potential fatalities who used both restraints. This includes motorists who would have lived with belts and no bags, and those who would have lived with either device. We attribute these 732 to belts and the remaining 138 (870 - 732) to bags.

Combining this with the survivors who only used a single restraint (or were under 13) yields the attribution in Table 11 of 1,440 lives saved among right-front passengers in cars with 3-point belts.

Table 11: Attribution for the Lives Saved Among Right-Front Passengers in Cars With 3-Point Belts in 2002

Lives Saved	
Seat Belts	**Air Bags**
1,159	281
Data derived from: NCSA, NHTSA, FARS, 2002	

The Calculation Nationwide

Applying the same calculation nationwide yields the following attribution in Table 12 of the 16,441 occupants who were saved by seat belts or air bags in 2002.

Table 12: Attribution for the Lives Saved by Seat Belts and Frontal Air Bags Nationwide in 2002	
Lives Saved	
Seat Belts	**Air Bags**
14,154	2,288
Data derived from: NCSA, NHTSA, FARS, 2002	

State Calculations

As with the lives saved by belts and/or bags, we calculate the attributions by applying the above formula to State fatality data, and normalizing the results to sum to the nationwide total. For example, we calculate the lives saved by seat belts in Alabama by applying the formula to all States, and multiplying the Alabama number by X/Y, where X is the lives saved by seat belts nationwide and Y is the total lives saved by seat belts obtained by applying the formula to each State.

7.   Potential Lives Saved

In order to underscore the importance of seat belt use, NHTSA estimates the numbers of lives savable in a given year (among passenger vehicle occupants 5 and older) if the national seat belt use rate had attained various higher values, such as 1 percentage point higher, 90 percent, or 100 percent. We derive these calculations in this section, first for one device, then two, and then we apply the two-device scenario to seat belts and air bags.

7.1   For a Sole Device

Suppose that more people use Device A. For example, say that a proportion of x_{actual} people actually used A, and we hypothesize x_{hypoth} to use A, where $x_{hypoth} > x_{actual}$. Suppose that the corresponding use rates among potential fatalities are u_{hypoth} and u_{actual}, respectively. Then an additional proportion of $u_{hypoth} - u_{actual}$ potential fatalities would have used A. If there are P potential fatalities, then

$$(e)(u_{hypoth} - u_{current})(P)$$

of the new users would have lived. That is, a total of $(e)(u_{hypoth})(P)$ people would have lived if the use rate had been x_{hypoth}, and the corresponding use rate among potential fatalities had been u_{hypoth}.

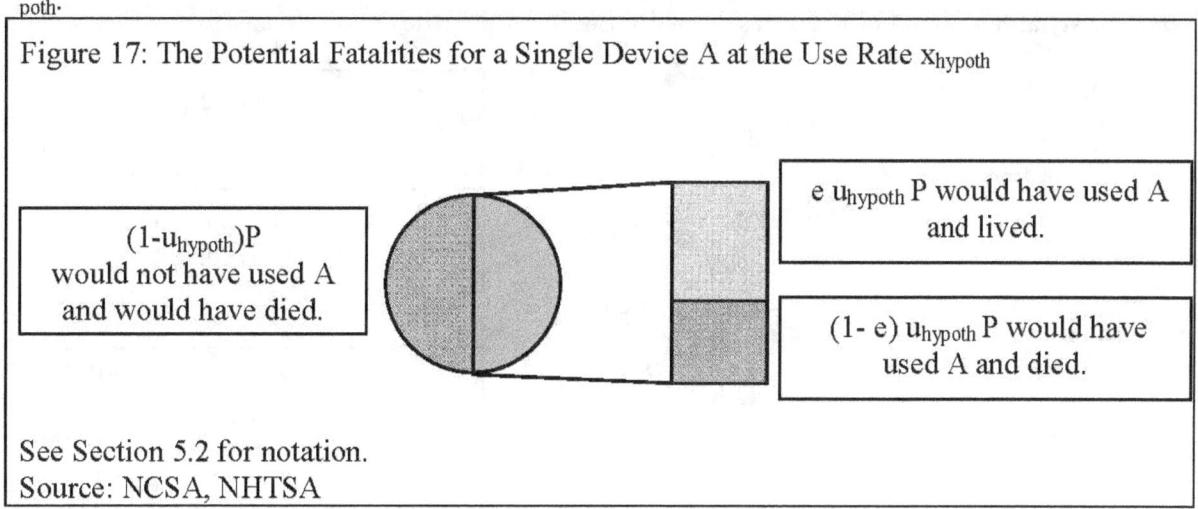

Figure 17: The Potential Fatalities for a Single Device A at the Use Rate x_{hypoth}

$(1-u_{hypoth})P$ would not have used A and would have died.

$e\ u_{hypoth}\ P$ would have used A and lived.

$(1-e)\ u_{hypoth}\ P$ would have used A and died.

See Section 5.2 for notation.
Source: NCSA, NHTSA

7.2   For Multiple Devices

The multiple-device scenario is significantly more complicated. In our application to seat belts and air bags, we will only consider the scenario in which seat belts are hypothesized to have a higher use rate, although one could also consider what would happen if air bags had been in a higher proportion of vehicles. Consequently, we will only consider the scenario in which one of the devices, say A, is hypothesized to have a higher use rate.

For reference, we will refer to the instances of the setting as they actually occurred, with A's actual use rate, as the Actual Scenario. We will refer to the instances of the setting, under which A is hypothesized to have a certain specified higher use rate, as the Hypothetical Scenario.

For instance, if A is taken to be seat belts in the two-device seat belt and air bag setting, then the Actual Scenario is the set of crashes in, say, some given year, as they actually occurred, with a belt use rate of, say, 79 percent. If we wish to consider what would happen if belt use had been 90 percent instead of 79 percent, then the Hypothetical Scenario would be a set of actual and modified crashes from the same year, in which we change the belt use of a random subset of motorists who had not used seat belts in order to make the overall belt use rate 90 percent.

7.2.1 Potential Lives Saved

Suppose that A is hypothesized to have the use rate $x_{hypoth} \times 100\%$, with the corresponding use rate among potential fatalities $u_{hypoth} \times 100\%$. Then an additional fraction of $u_{hypoth} - u_{actual}$ potential fatalities would have used A, where u_{actual} is the current use rate among the potential fatalities.

We first recall, using the notation of Section 5.2, how the potential fatalities actually broke down, in terms of which device(s) they used.

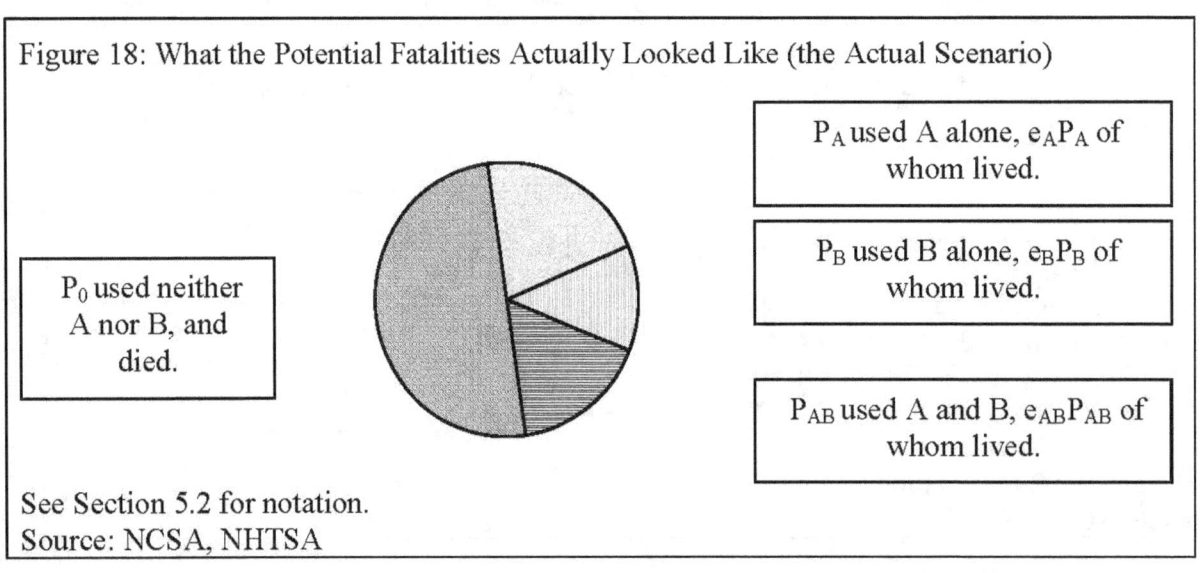

Figure 18: What the Potential Fatalities Actually Looked Like (the Actual Scenario)

P_A used A alone, $e_A P_A$ of whom lived.

P_B used B alone, $e_B P_B$ of whom lived.

P_0 used neither A nor B, and died.

P_{AB} used A and B, $e_{AB} P_{AB}$ of whom lived.

See Section 5.2 for notation.
Source: NCSA, NHTSA

All potential fatalities who used A (or used A and B) continue to do so in the hypothetical scenario. In addition, a random subset of $(u_{hypoth} - u_{actual})P$ potential fatalities are chosen from those who did not use A, to use A in the Hypothetical Scenario. In the equations below, N_{AB} potential fatalities are (hypothetically) newly using A and B, having previously used only B, while N_A potential fatalities are (hypothetically) newly using A, having previously used neither device. In the equations below, P represents the total number of potential fatalities, P_B represents the potential fatalities who used B alone, and P_0 represents the potential fatalities who used neither Device A or Device B.

In this random subset, we would expect to encounter

$$N_{AB} := (u_{hypoth}-u_{actual})PP_B/(P_B+P_0)$$

who had actually used only B, and

$$N_A := (u_{hypoth}-u_{actual})PP_0/(P_B+P_0)$$

who had actually used neither device.

That is, in the Hypothetical Scenario, $\underline{N_{AB}\ potential\ fatalities}$ are (hypothetically) newly using A and B, having previously used only B, while $\underline{N_A\ potential\ fatalities}$ are (hypothetically) newly using A, having previously used neither device. The potential fatalities as hypothesized appear in Figure 19, with the above defined N_A and N_{AB} included in Figure 19.

Figure 19: What the Potential Fatalities Would Have Looked Like if x_{hypoth} Had Used A (the Hypothetical Scenario)

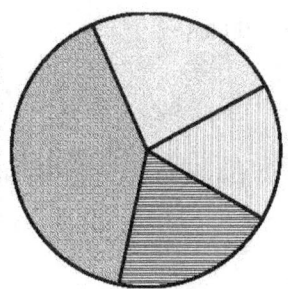

$P_0 - N_A$ would have used neither A or B, and would have died.

$P_A + N_A$ would have used A alone, $e_A (P_A+N_A)$ of whom would have lived.

$P_B - N_{AB}$ would have used B alone, $e_B (P_B-N_{AB})$ of whom would have lived.

$P_{AB} + N_{AB}$ would have used A and B, $e_{AB} (P_{AB}+N_{AB})$ of whom would have lived.

See Section 5.2 for notation.
Source: NCSA, NHTSA

In the Actual Scenario, a total of $e_A P_A+e_B P_B+e_{AB} P_{AB}$ potential fatalities lived. In the Hypothetical Scenario, a total of

$$e_A (P_A+N_A) + e_B (P_B-N_{AB}) + e_{AB} (P_{AB}+N_{AB})$$

lived. These are depicted in Figure 20, on the following page.

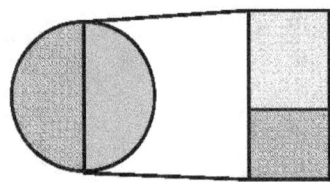
The net increase in saved lives when one goes from the actual to the hypothetical scenarios is

$$e_A N_A + (e_{AB} - e_B)N_{AB}.$$

The first term represents the people who actually used neither device but were saved by A in the Hypothetical Scenario. The second term represents the people who actually died using B alone, but were saved when they added A in the Hypothetical Scenario.

The additional lives saved can alternatively be expressed as

$$(u_{hypoth} - u_{actual}) \, P \, (e_A P_0 + (e_{AB} - e_B)P_B)/(P_0 + P_B).$$

One can obtain formulas for the attributions in terms of fatalities by substituting the relationships, such as $P_A = F_A/(1 - e_A)$, from Section 4.2.4.

7.3   For Seat Belts and Frontal Air Bags

In this section we determine how many lives would have been saved if seat belt use had been at a higher rate x_{hypoth} ($\times$ 100%). Although we do not do so, the same method could be used to determine the potential lives saved if more vehicles had air bags.

Recall that the two-device scenario first requires determining what the belt use rate u_{hypoth} among the potential fatalities will be when the use rate in the general population is x_{hypoth}. We do this in the next section. The subsequent section then applies the two-device formulas from Section 7.2.

Daytime Front Seat Use

We shall see in the next section that, because of available data on belt use, our calculations will not quite produce the lives savable if belt use had been at a higher rate x_{hypoth}, but rather those saved if belt use <u>in the front outboard seats in daytime</u> had been x_{hypoth}. (The front outboard seats are the driver's seat and that of the right-front passenger.) However, the qualification "front outboard seats during daytime" is frequently dropped when communicating the results, as we will often do in this report.

Predicting Future Lives Saved

The formulas in this section calculate the lives savable in a particular year if belt use had been higher. NHTSA also frequently estimates the lives savable in the future if belt use reaches a specified use rate in the future, e.g., when estimating the potential impact of passing stronger (primary) belt laws. This is often done without specifying a point in the future at which one expects belt use to reach the hypothesized rate.

Future lives saved are frequently predicted using the formulas in this section, estimating the lives savable if belt use reaches x_{hypoth} in some future year as the lives savable in the most recent year for which we have fatality data if use had been x_{hypoth}. This results in a conservative estimate: People have spent more time and miles on the road each year, and so there should be more potential fatalities, and more savable lives, in the future than there are now.[1]

If a future year is specified during which we expect belt use to reach the hypothesized rate, a more accurate estimate of the savable lives would be obtained by multiplying the conservative estimate by the expected increase in vehicle miles traveled (VMT). VMT has been growing by approximately 2 percent each year.

Attribution in the Hypothetical Scenario

We will keep the same attribution of the lives saved (i.e., we will not change the attribution of anyone who was saved), <u>and attribute all additional lives saved to seat belts</u>. Recall that no additional vehicles are hypothesized to have air bags under our Hypothetical Scenario.

7.3.1 Determining the Hypothetical Use Among Potential Fatalities

Recall that the first step in computing potential lives saved is to determine the belt use rates u_{hypoth} and u_{actual} among potential fatalities when use in the general population is x_{hypoth} and x_{actual}. We would expect belt use to be lower in the former group since it contains a greater number of "risk takers." We accomplish the task by using a model that relates belt use in these two populations.

---

[1] We note however that VMT has declined in recent years, which is an unusual occurrence.

We do not have data on belt use in all seating positions and times of day. Our best data comes from surveys that observe belt use in the front outboard seating positions during daylight hours. Consequently we cannot derive a good model predicting use among potential fatalities from use in the general population. The models NHTSA uses instead predict use among potential fatalities from use in the front outboard seats during daytime.

We will refer to belt use among potential fatalities as "UPF." Recall that potential fatalities are those occupants, in a given data year, in crashes in which they would have died if they had been unbelted and hadn't had a frontal air bag. The most recent model of UPF, as published in Wang and Blincoe (2003), is

$$y = 0.47249 \, x^2 + 0.43751 \, x,$$

where y denotes UPF and x denotes front seat daytime use. This model has an adjusted R-squared of 0.9941. Its predictions for the values 75% and higher are given in Table 13. Note that, as expected, the predicted values (y) are smaller than the general use (x). For instance, when national (daytime front seat) use is 100%, we expect that 9 percent (100-91=9) of potential fatalities are still not using belts.

Daytime Front Seat Use
Because the model from Wang and Blincoe (2003) takes belt use in the front seat during daytime as input, our lives savable cal-

Table 13: Belt Use Among Potential Fatalities to Daytime Front Seat Use in the General Population	
Belt Use in the United States in the Front Seat During Daytime x	Belt Use Among Potential Fatalities in the United States Predicted from the Model UPF(x) = $0.47249 \, x^2 + 0.43751 \, x$
75%	59%
76%	61%
77%	62%
78%	63%
79%	64%
80%	65%
81%	66%
82%	68%
83%	69%
84%	70%
85%	71%
86%	73%
87%	74%
88%	75%
89%	76%
90%	78%
91%	79%
92%	80%
93%	82%
94%	83%
95%	84%
96%	86%
97%	87%
98%	88%
99%	90%
100%	91%
Source: (Wang and Blincoe, 2003)	

culations actually determine the lives savable if daytime front seat use had been x_{hypoth}, not if general use had been x_{hypoth}. However, the resulting estimates are frequently referred to as the lives saved if belt use had been, e.g., 90 percent or 100 percent (as opposed to front seat daytime use being 90% or 100%).

Applying the Model Nationwide
When estimating the savable lives in individual States, there is an obvious source for x_{actual}, namely State belt surveys (Glassbrenner, May 2003). These are probability-based observational surveys that provide the best estimates of belt use at the State level.

However when estimating the savable lives nationwide, there are two possible estimates of national use to which the UPF model could be applied. NHTSA measures daytime front seat use nationwide in its National Occupant Protection Use Survey (NOPUS) (Glassbrenner, September 2003). Like the State surveys, NOPUS is probability-based and observes belt use as it actually occurs on the road. We will call its estimate of nationwide use the *NOPUS estimate*. On the other hand, State surveys could be combined (typically weighted according to traffic volume) into a national estimate, which we will call the *State-based national estimate, SBNE*.

The SBNE has been consistently higher than the NOPUS estimate. This reflects cost saving measures (such as the exclusion of a percentage of rural areas from the sampling frame), employed by nearly all States but not by NOPUS, that result in upward biases (Glassbrenner, May 2003). The net bias at the national level can vary from year to year as States change their protocols and as their traffic volumes (and hence their contributions to the national estimate) fluctuate. On average the SBNE has been about 2 percentage points higher than the NOPUS Estimate. In this subsection we explain why we will apply the UPF model to the NOPUS estimate instead of the SBNE.

The model from Wang and Blincoe (2003) was fitted with belt use rates from State surveys and State fatality data. This would seem to argue that the model reflects State protocols and so should be applied to the SBNE, instead of NOPUS. However the model was fit largely using surveys prior to 1998, and these were very different from surveys conducted today. Starting in 1998, nearly all States followed a set of criteria established by NHTSA to ensure a certain degree of uniformity. Because of the criteria, surveys conducted after 1997 were probability-based and observed all vehicle and motorist classes. Many earlier surveys were conducted on convenience samples and/or only observed the vehicles and motorists covered under their belt law at the time. For instance, some States did not observe pickup trucks or passengers in the right-front seat, and this can have an appreciable effect on the State's estimate. Thus the survey protocols prior to 1998 differed to a substantial degree from today's surveys, and the argument that the model should be applied to the SBNE because the model reflects State protocols is greatly diminished.

NOPUS provides a more stable national estimate than the State surveys. While the NHTSA criteria for State surveys ensure some degree of uniformity, substantial differences in the design from State to State remain. These can include nontrivial differences in sample design, observation protocols, and estimation procedures, and States can change their procedures at any time.

Finally, State survey results are not available at the time NHTSA conducts its calculation of savable lives. NOPUS has a quick turnaround, producing estimates a few months after the survey is conducted. States do not need to report their results until March of the calendar year following the data year. Using State surveys would substantially delay NHTSA's estimation of savable lives. As a consequence of all the above reasons (stability, availability, and the protocols reflected by the UPF model), we will apply the UPF model to the NOPUS estimate of national belt use when calculating the savable lives nationwide.

In subsequent sections, "UPF" will denote the function from the above model. That is, UPF(x) := $0.47249\ x^2 + 0.43751\ x$.

7.3.2 Potential Lives Saved

As we did with lives saved, we apply the two-device formulas from Section 7.2 to the setting of seat belts and air bags. This will give us formulas for the lives saved by belts and/or bags, and for the additional saved (which are all attributed to belts). We illustrate each on Example 1 and give the nationwide figures.

Deriving the Formula

Recall that our two-device formulas can either be applied for each belt type and summed over the belt types or by applying an average belt type (with no summing). The latter approach is generally less desirable because it is susceptible to errors in estimating the average belt effectiveness, which would change each year as vehicles on the road do. We used the former approach to calculate lives saved, but this approach would produce substantially more complex formulas for lives savable. We will use a hybrid procedure that estimates some terms via the simpler and some via the complex approaches. This will produce more practicable formulas, without much loss of accuracy. In the following we will refer to the two approaches as the average and detailed approaches.

Recall that the two-device formula for the additional lives saved at use x_{hypoth} is

$$e_A N_A + (e_{AB} - e_B)N_{AB} \qquad (3)$$

where $N_A = (u_{hypoth}-u_{actual})PP_0/(P_B+P_0)$ and $N_{AB} = (u_{hypoth}-u_{actual})PP_B/(P_B+P_0)$. We have the following estimates of potential fatality terms, using the detailed approach.

$$P = \sum_{i \in R} \frac{F_i}{1 - e_i \text{ (used)}}, \; P_0 = \sum_{\substack{belt(i)=0 \\ bag(i)=0}} F_i \; , \; P_B = \frac{1}{1 - e(bag)} \sum_{\substack{belt(i)=0 \\ bag(i)=1}} F_i \; ,$$

while of course $u_{hypoth} = UPF(x_{hypoth})$, $u_{actual} = UPF(x_{actual})$, and $e_B = e(bag) = 0.14$. Here "UPF" denotes the function from the model in the last section, namely $UPF(x) = 0.47249 \; x^2 + 0.43751 \; x$.

We will estimate e_A and e_{AB} using the average approach, taking A to be an average belt and B to be an air bag. Recall that in (3), these effectiveness ratings are applied to people who in the Actual Scenario do not use A. Specifically e_A is applied to potential fatalities who actually use neither A nor B, and e_{AB} to those who actually use B alone. Thus, rather than using average effectiveness ratings among <u>all</u> potential fatalities, it would be more accurate to estimate as e_A as the average belt rating among unbelted potential fatalities without air bags, and e_{AB} as the average rating of the belt-bag system among unbelted potential fatalities with air bags. These will differ from the averages among <u>all</u> potential fatalities because, e.g., automatic belts are nearly always in use and have a relatively low effectiveness rating. Consequently, we will take

$$e_A := \sum_{\substack{belt(i)=0 \\ bag(i)=0}} e_i \text{ (belt) } F_i \; \bigg/ \sum_{\substack{belt(i)=0 \\ bag(i)=0}} F_i \; , \text{ and}$$

$$e_{AB} := \sum_{\substack{belt(i)=0 \\ bag(i)=1}} e_i\,(\text{system})\,F_i \Bigg/ \sum_{\substack{belt(i)=0 \\ bag(i)=1}} F_i \ .$$

Note that here, averaging over the potential fatalities is equivalent to averaging over the fatalities. This occurred because in each case, the potential fatalities over which we are averaging all have the same effectiveness (0% for the potential fatalities used for e_A and 14% for e_{AB}).

With these choices, (3) becomes

$$e_A N_A + (e_{AB} - e_B)N_{AB} =$$

$$\left(\sum_{\substack{belt(i)=0 \\ bag(i)=0}} e_i\,(\text{belt})\,F_i \Bigg/ \sum_{\substack{belt(i)=0 \\ bag(i)=0}} F_i\right)\left(UPF(x_{hypoth}) - UPF(x_{actual})\right)\sum_{i \in R}\frac{F_i}{1 - e_i\,(\text{used})} \sum_{\substack{belt(i)=0 \\ bag(i)=0}} F_i \Bigg/ \sum_{belt(i)=0}\frac{F_i}{1 - e_i\,(\text{used})} \ +$$

$$\left(\left(\sum_{\substack{belt(i)=0 \\ bag(i)=1}} e_i\,(\text{system})\,F_i \Bigg/ \sum_{\substack{belt(i)=0 \\ bag(i)=1}} F_i\right) - e(\text{bag})\right)\left(UPF(x_{hypoth}) - UPF(x_{actual})\right)\sum_{i \in R}\frac{F_i}{1 - e_i\,(\text{used})} \sum_{\substack{belt(i)=0 \\ bag(i)=1}}\frac{F_i}{1 - e(\text{bag})} \Bigg/ \sum_{belt(i)=0}\frac{F_i}{1 - e_i\,(\text{used})}$$

In this formula one can see how the calculation is arrived at. The first term estimates the proportion of newly buckled who do not have bags, and applies their average belt effectiveness rating to determine who among them is saved. The second term estimates the proportion of newly buckled who have air bags, applies their average belt-bag effectiveness rating to determine who among them is saved, and subtracts out those who were saved by their air bags.

Simplifying, the number of additional lives savable if seat belt use had been x_{hypoth} can also be expressed as

$$\left(UPF(x_{hypoth}) - UPF(x_{actual})\right)\left(\sum_{\substack{belt(i)=0 \\ bag(i)=0}} e_i\,(\text{belt})\,F_i + \sum_{\substack{belt(i)=0 \\ bag(i)=1}}\left(e_i\,(\text{system}) - \frac{e(\text{bag})}{1 - e(\text{bag})}\right)F_i\right)\sum_{i \in R}\frac{F_i}{1 - e_i\,(\text{used})} \Bigg/ \sum_{belt(i)=0}\frac{F_i}{1 - e_i\,(\text{used})}$$

Notes on the Derivation

Recall that one could have applied the two-device formula to each type of seat belt to determine the additional lives saved if the appropriate number of unbelted with that belt type buckled up, and then summing to get the additional saved among all belt types. However we have seen in practice that the increased accuracy that comes from this is not worth the additional complexity.

Note that since the UPF model uses belt use in the front seat of passenger vehicles in the daytime, the above formula really calculates the additional lives saved if front seat daytime use had been x_{hypoth}. However the qualification "front seat daytime" is frequently dropped for communicability.

Note that we are **not** computing the additional lives saved by estimating the total lives saved in the natural way from the UPF model and subtracting the current lives saved. This would have resulted in a discontinuity, since the UPF model does not precisely predict the use rate that actually occurred among the potential fatalities when the actual daytime front seat use is plugged in to it.

Formula
We have calculated that the number of additional lives savable if seat belt use had been x_{hypoth} is

$$\left(UPF(x_{hypoth}) - UPF(x_{actual})\right) \left(\sum_{\substack{belt(i)=0 \\ bag(i)=0}} e_i(belt) F_i + \sum_{\substack{belt(i)=0 \\ bag(i)=1}} \left(e_i(system) - \frac{e(bag)}{1 - e(bag)} \right) F_i \right) \sum_{i \in R} \frac{F_i}{1 - e_i(used)} \Bigg/ \sum_{belt(i)=0} \frac{F_i}{1 - e_i(used)}$$

Figure 21: The Potential Fatalities if Belt Use Had Been x_{hypoth}

$$\sum_{i \in R} \frac{e_i(used) F_i}{1 - e_i(used)} +$$

$$\left(UPF(x_{hypoth}) - UPF(x_{actual})\right) \left(\sum_{\substack{belt(i)=0 \\ bag(i)=0}} e_i(belt) F_i + \sum_{\substack{belt(i)=0 \\ bag(i)=1}} \left(e_i(system) - \frac{e(bag)}{1 - e(bag)} \right) F_i \right) \sum_{i \in R} \frac{F_i}{1 - e_i(used)} \Bigg/ \sum_{belt(i)=0} \frac{F_i}{1 - e_i(used)}$$

had air bags or would have used belts, and lived.

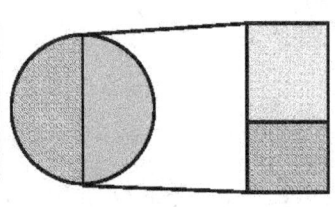

Unbelted potential fatalities without air bags and unbelted children 5 to 12 would have died.

The remaining potential fatalities would have used seat belts or had air bags, and died.

See Section 5.2 for notation.
Source: NCSA, NHTSA

To use a perhaps simpler notation, the formula for the additional lives saved is:

$$(e(belt) \, P_{unrest} + (e(system) - e(bag)) \, P_{bag}) \, (\Delta u) \, P/(P_{unrest} + P_{bag})$$

44

where $\Delta u := UPF(x_{hypoth}) - UPF(x_{actual})$ is the change in UPF, $P := \sum_{i \in R} \dfrac{F_i}{1 - e_i(\text{used})}$ is the number

of potential fatalities, $P_{unrest} := \sum_{\substack{belt(i)=0 \\ bag(i)=0}} F_i$ is the number of unrestrained potential fatalities, $P_{bag} :=$

$\dfrac{1}{1 - e(\text{bag})} \sum_{\substack{belt(i)=0 \\ bag(i)=0}} F_i$ is the number of unbelted potential fatalities with air bags, $e(\text{belt}) :=$

$\sum_{\substack{belt(i)=0 \\ bag(i)=0}} e_i(\text{belt}) F_i \Bigg/ \sum_{\substack{belt(i)=0 \\ bag(i)=0}} F_i$ is the average belt effectiveness (among the unrestrained), and $e(\text{system})$

$:= \sum_{\substack{belt(i)=0 \\ bag(i)=1}} e_i(\text{system}) F_i \Bigg/ \sum_{\substack{belt(i)=0 \\ bag(i)=1}} F_i$ is the average system effectiveness (among the unbelted with bags).

Example 1

We calculate the additional right-front passengers in cars with 3-point belts in 2002 who would have been saved if (daytime front seat) belt use had been 90% in 2002. The UPF model from (Wang

Table 14: The Newly Buckled Right-Front Passengers in Cars With 3-Point Belts in Potentially Fatal Crashes in 2002 if Belt Use Nationwide Had Been 90 Percent

Age 5-12?	Belt Used?	Airbag Present?	Effec-tiveness	Potential Fatalities	Lives Saved	Lives Previously Saved by Bags	Net Lives Saved
Yes	Yes	NA	37%	12	5	0	5
No	Yes	Yes	44%	498	219	70	149
No	Yes	No	37%	406	150	0	150
Totals*				917	374	70	304

*Items do not necessarily sum to totals, due to rounding.
Source: NCSA, NHTSA, FARS, 2002

and Blincoe, (2003) estimates that UPF would have been 18 percentage points higher in 2001 if (daytime front seat) use had been 90 percent. (Belt use (in the front seat in daytime) was 75 percent in 2002. The model estimates 78 percent UPF at 90 percent use and 59 percent at 75 percent use, an increase of 19 percentage points, or 18, if one uses all decimal places.) That is, the UPF model predicts that 18 percent more of the potential fatalities would have buckled up if daytime front seat use had been 90 percent.

Applying this to the 5,021 potential fatalities gives that 917 additional potential fatalities would have buckled up (again using all decimal places).

Table 14 computes the lives saved among the newly buckled according to the techniques used in Section 6. We illustrate the calculation for the occupants over 12 who had an air bag.

First we calculate how many of the 917 newly buckled will fit this description (over age 12 and have an air bag). Referencing Table 10 we see that in the crashes they actually experienced,

1,026 of the 1,888 unbelted potential fatalities were over 12 and had an air bag. So in the group of 917 randomly selected to buckle, we would expect 498 of them to be over 12 and have an air bag (i.e., 498 = 917 × 1,026/1,888).

Applying the 44 percent effectiveness of the belt-bag combination, we see that 219 of these newly buckled (the 498 newly buckled over age 12 with bags) will live. Applying the 14 percent effectiveness of the air bag, we see that 70 lived when they were unbelted (i.e., 70 is 14% of the 498 newly buckled). So an additional 149 right-front passengers over age 12 in cars with air bags and 3-point belts would have lived in 2002 if (daytime front seat) belt use had been 90 percent.

Similarly, we see that 304 additional lives would have been saved in passenger cars with 3-point belts in 2002 if belt use had been 90 percent.

The actual UPF in 2002, calculated from FARS, was 42 percent. Note however that we are using the predicted value of 59 percent when estimating the increase in UPF resulting from 90 percent use. We do this for consistency. The model obviously has some error, and it is reasonable to think that since its predicted current UPF (of 59%) was an overestimate, its predicted hypothetical UPF (at 90% use) is likely to overestimate as well. So it is better to estimate the change in UPF using the predicted value of 59 percent rather than the actual value of 42 percent as the estimate of current UPF.

The Calculation Nationwide
Applying this calculation nationwide gives the following numbers for the 2002 data year:

- 260 additional lives would have been saved if (front seat daytime) belt use had been one point higher (76%);
- 4,116 if use had been 90 percent; and
- 7,127 if use had been 100 percent.

State Calculations
Calculations at the State level (e.g., the lives saavable in Alabama in 2002 if its seat belt use had been 100%) are obtained by applying the formulas from this section to every State using State fatality data and State belt use rates. These preliminary estimates are then normalized to sum to the nationwide total lives lost calculated with the formulas from this section.

Lives Saved at 100 Percent Belt Use
Recall that our formula for the lives savable if seat belt use had been x_{hypoth} actually calculates the number of lives savable if daytime front seat use had been x_{hypoth}. This was necessary because we needed to relate use among the potential fatalities to use among the general population use, and the only data we have on the latter is daytime front seat use.

However for one hypothesized use rate, namely 100 percent use, we could also calculate the number of lives savable if the general population use had been 100 percent (literally, if everyone buckled up), as

$$\sum_{i \in R} \frac{e_i\,(\text{system})\,F_i}{1 - e_i\,(\text{used})}$$

This formula produces a substantially larger number than that from our formula for lives savable (i.e., the formula preceding Figure 20) when $x_{hypoth} = 1$. (For example the formula preceding Figure 20 estimates about 7,000 lives in 2002, while the above formula estimates about 9,000.) If we used the formula preceding Figure 20 to calculate the savable lives when $x_{hypoth} < 1$ and that above when $x_{hypoth} = 1$, we would have a discontinuity, and an unreasonable increase, at $x_{hypoth} = 1$. Consequently we use the formula preceding Figure 20 to calculate the savable lives for all values of x_{hypoth}. For communicability, however, NHTSA publications, such as (NCSA's *Traffic Safety Facts 2007 - Occupant Protection* (undated), and speeches by the NHTSA Administrator describe the lives saved if daytime front seat use had been 100 percent as the lives saved if use had been 100 percent.

8. Summary of Formulas

The following table summarizes the formulas presented in this report.

Table 15: Formulas Concerning Lives Saved, and Potential Lives Saved, by Seat Belts and Frontal Air Bags

Item	Formula
Potential fatalities	$$\sum_i \frac{F_i}{1-e_i(\text{used})}$$
Lives saved by belts and/or bags	$$\sum_i \frac{e_i(\text{used})\,F_i}{1-e_i(\text{used})}$$
Lives saved by seat belts	$$\sum_{\text{belt}(i)=1} \frac{e_i(\text{belt})\,F_i}{1-e_i(\text{used})}$$
Lives saved by air bags	$$\frac{e(\text{bag}\,|\,\text{belt})}{1-e(\text{bag}\,|\,\text{belt})}\sum_{\substack{\text{belt}(i)=1,\\\text{bag}(i)=1}} F_i + \frac{e(\text{bag})}{1-e(\text{bag})}\sum_{\substack{\text{belt}(i)=0,\\\text{bag}(i)=1}} F_i$$
Belt use among potential fatalities when (daytime front seat) use in the general population is x.	$$UPF(x) := 0.47249\,x^2 + 0.43751\,x$$
Additional lives savable if (daytime front seat) belt use had been x_{hypoth}*	$$\left(UPF(x_{\text{hypoth}})-UPF(x_{\text{actual}})\right)\left(\sum_{\substack{\text{belt}(i)=0\\\text{bag}(i)=0}} e_i(\text{belt})F_i + \sum_{\substack{\text{belt}(i)=0\\\text{bag}(i)=1}}\left(e_i(\text{system})-\frac{\text{bag}(i)e(\text{bag})}{1-e(\text{bag})}\right)F_i + \sum_{i\in R}\frac{F_i}{1-e_i(\text{used})}\right)\bigg/ \sum_{\text{belt}(i)=0}\frac{F_i}{1-e_i(\text{used})}$$
Additional lives saved if everyone had buckled up*	$$\sum_{i\in R}\frac{e_i(\text{system})F_i}{1-e_i(\text{used})} - \sum_{\text{belt}(i)=1}\frac{e_i(\text{belt})F_i}{1-e_i(\text{used})}$$
Lives savable by belts if belt use had been x_{hypoth}	$$\sum_{\text{belt}(i)=1}\frac{e_i(\text{belt})F_i}{1-e_i(\text{used})} + $$

Table 15: Formulas Concerning Lives Saved, and Potential Lives Saved, by Seat Belts and Frontal Air Bags

Item	Formula
	$$\left(UPF(x_{hypoth}) - UPF(x_{actual})\right)\left[\sum_{\substack{belt(i)=0\\bag(i)=0}} e_i(belt)\,F_i + \sum_{\substack{belt(i)=0\\bag(i)=1}}\left(e_i(system) - \frac{e(bag)}{1-e(bag)}\right)F_i\right]\left.\sum_{i\in R}\frac{F_i}{1-e_i(used)}\middle/\sum_{belt(i)=0}\frac{F_i}{1-e_i(used)}\right.$$
Lives savable by bags if belt use had been x_{hypoth}	$$\frac{e(bag|belt)}{1-e(bag|belt)}\sum_{\substack{belt(i)=1,\\bag(i)=1}}F_i + \frac{e(bag)}{1-e(bag)}\sum_{\substack{belt(i)=0,\\bag(i)=1}}F_i$$
* Unless otherwise specified, NHTSA literature uses the lives saved if daytime front seat use had been 100 percent for calculations of lives saved if use had been 100 percent. See Section 5.2 for notation.	

Source: NCSA, NHTSA

Appendix: The Treatment of Factors Affecting Survival as Devices or Part of the Setting

Usually a large (if not arguably innumerable) number of factors affect survival in a potentially life-threatening situation. Factors such as weather conditions, driver fatigue, ambient traffic, road design, the use of seat belts, and presence of air bags and side door beams are some of the perhaps innumerable factors affecting survival in vehicle crashes.

As we have mentioned, in order to estimate saved lives, one needs to decide for each factor whether the treat the factor as a device or whether to treat it as part of the setting. This determination is crucial, because the choice of treatments can substantially impact who shall be said to be saved, and by what device.

This section presents a variety of information concerning the decision to treat factors as devices versus part of the setting. In Appendix Section 1.1 (A1.1) we illustrate how an arguably unlimited number of factors can be said to affect survival in real world settings. Also, in order to use a factor as a device, one needs to have certain effectiveness ratings. In fact, one needs one type of effectiveness ratings to compute the total lives saved by all devices combined, and perhaps additional ratings to allocate the total to the savings credited to each device individually. We explain the ratings needed in Appendix Section 1.2 (A1.2).

The impact of considering a factor to be a device or part of the setting can be difficult to assess. Appendix Section 1.3 (A1.3) contains crucial analyses that illustrate how the choice of which factors to treat as devices affects who is said to be saved, and by what device.

In Appendix Section 1.4 (A1.4) we put all of this information together to illuminate the choice of which factors should be treated as devices when computing lives saved. After reading the examples in Appendix Section 1.3 (A1.3), it is perhaps best to decide treatment based on (1) whether the factor has been rated for effectiveness, and (2) philosophical views on whether the factor is viewed as part of the setting or as a device acting in the setting. We encourage readers to investigate the consequence of their choice by conducting analyses like those in Appendix Section 1.3 (A1.3) before implementing them in their calculations.

A1.1 A Virtually Unlimited Number of Factors Could Be Argued to Affect Survival.

Few scenarios are as pure as the single-device scenario. Settings such as car crashes are complex phenomena, typically with several factors such as belts, crumple zones, side-door beams, guardrails, weather, road design, health, and quick thinking playing roles in survival. Following is a typical example.

Example 1

A physically fit woman driving a 2000 Saab 9-5 sedan swerves on a rainy undivided highway to avoid colliding with a truck that, through loss of control, has crossed into oncoming traffic. The woman's Saab crashes into a guardrail, her air bag systems (a frontal and side bag) deploy, and she lives. The woman was belted. Had it not been raining, she would have avoided both the truck and the guardrail, and the crash would not have occurred. Had there been ice on the highway, the Saab would have spun off a section of the roadway not protected by a guardrail and the woman would have died. Other than her Saab's crumple zone, none of the passive systems in her vehicle (including the air bags, side door beams, etc.) played a role in her survival. Had she not been physically fit, she would not have survived her injuries. In addition she needed either her belt or a crumple zone to live but not both. Had there not been a guardrail where her Saab departed the roadway, she would have crashed head-on into a tree and died.

To what, if anything, would we attribute this woman's survival? Most people would say that the guardrail, seat belt, crumple zone, and the woman's physical fitness all played roles in survival. Without further information, the factors cannot be ranked in order of their contributions, and in particular no single factor can be isolated as the key element in survival.

However, not all people would end the list here, with factors that contribute to the circumstances of the crash being particularly debatable. Some might point to the fact that the woman would have died if there had been ice or had she not had the presence of mind to swerve, and conclude that weather and the driver's reaction also contributed to survival. Others might say that these elements helped determine whether and how the crash occurred, but did not affect survival once the crash was underway. Devices that are designed to help prevent crashes from happening, such as daytime running lamps, are similarly debatable. In this paper we take the view that devices that affect whether or how a crash occurs, but do not affect survival once the crash is underway, do not contribute to survival.

There is no canonical answer to what saved someone. There are typically an unlimited number of factors that affect survival in some fashion, and people can reasonably argue about which factors are reasonably considered to contribute to survival, not to mention which if any, was the principal contributor.

What we do have at our disposal is information on nearly all traffic fatalities (those that occur within 30 days of a crash of a vehicle). As we discuss in the next section, one can estimate from this database the number of crash survivors whose survival was influenced by a specified set of factors if one has effectiveness ratings (rating the effectiveness against fatality) for these factors.

A1.2 Devices Must Be Rated for Effectiveness.

Each factor that is considered as a device must be rated for its effectiveness in preventing fatality in order to be able to calculate the lives saved by the devices. For instance, NHTSA does not currently estimate the lives saved due to driver alertness, because this factor has not been rated for effectiveness. Indeed it seems unlikely that one could quantify an effectiveness rating for this type of amorphous factor.

Certain ratings are needed to compute the total lives saved by all devices, and, depending on how you decide to attribute survival to individual devices, additional ratings may also be required.

A1.2.1   Ratings Needed to Compute the Combined Lives Saved by All Devices

In order to determine the total lives saved by all devices combined, one must have the fatality-reducing effectiveness of all combinations of devices that could feasibly occur in an instance of the setting. Oftentimes the devices represent technologies that were implemented at various times. In this case, it suffices to have the effectiveness of each device in the presence of all previous devices, plus the variations on this in which previous devices that require active participation are not used.

For instance, NHTSA has rated each new vehicle technology required by the Federal Motor Vehicle Safety Standards for their effectiveness at preventing fatality for motorists in vehicles in which all previous FMVSS were implemented (FMVSS 208, 1996). Indeed this computation was required by OMB Circular A-4 for the benefits analysis of each proposed new FMVSS (OMB, 2003) Using these ratings one can compute the total lives saved by all FMVSS technologies, as was done in Kahane (2004).

Effectiveness ratings may be impractical to obtain. Many safety measures have not, or cannot reasonably, be rated for effectiveness. The effects of factors such as weather and driver reactions would seem to be highly variable and difficult to quantify. Such factors cannot be considered devices (i.e., must be considered part of the setting), and so we cannot attribute survival to these factors from information on fatal crashes.

A1.3   The Effect of Considering an Air Bag as a Device Versus Part of the Vehicle Crash

Suppose for simplicity that we were interested in estimating the lives saved by seat belts, not those saved by belts and bags. Consider the following two choices of devices.

In Choice 1, we choose seat belts to be the sole device. As we saw in Chapter 6, people are saved by the belts in Choice 1 if they used the seat belts, survived the crashes, and would have died if they hadn't used their seat belts. That is, they are the belted survivors who needed the seat belts to live.

In Choice 2, we choose belts and bags to comprise the devices. Using the attribution from Chapter 6, people are saved by the belts in Choice 2 if they used the belts, survived, would have lived without the air bags, and would have died with neither the belts nor bags.

Table 16: Who Is Saved Depends on the Choice of Devices

The Devices	Who Is Saved by Belts Under This Choice of Devices
Seat Belt	Belted survivors who need the belt to live.
Seat Belt, Air Bag	Belted survivors who would die without the belt and bag, and would live without the air bag.

Both Choice 1 and Choice 2 attribute to belts those survivors for whom the belt played the sole role in survival. These include people who used the belt but not the bag and survive, those who use the belt and bag but did not need the bag. Both choices also attribute belts with those whose survival depended on their belt and other elements not listed among the devices (e.g., survivors who required both their belt and crumple zone to live). That is, Choices 1 and 2 agree on the attribution of the easy cases (those for whom the belt played the sole role in survival) and in some of the hard cases (those for whom the belt contributed to, but was not solely responsible for, survival).

Where Choices 1 and 2 differ is the treatment of people for whom both belts and bags played roles in survival. A person who needs both the belt and bag to live is saved by the belt under Choice 1 but not Choice 2. A person who would live with either the belt or the bag is saved by the belt under Choice 2 but not Choice 1. Likewise a person who needs the belt, bag, and crumple zone is saved by the belt in Choice 1 but not Choice 2, while a person who needs the crumple zone and either the belt or bag is saved by the belt in Choice 2 but not Choice 1.

Table 17: What Saved Someone Depends on the Choice of Devices

Type of Survivor	Is the Survivor Saved by the Seat Belt When the Following Are Considered the Devices?	
	Seat Belt	Seat Belt, Air Bag
A belted survivor who needs both the belt and bag to live.	Yes	No
A belted survivor who would have lived with either the belt or the bag.	No	Yes
A belted survivor who needs the belt, bag, and crumple zone to live.	Yes	No
A belted survivor who needs the crumple zone and either the belt or bag to live.	No	Yes

As we can see, each choice is reasonable. Each counts the clear cases as saved by the belt (those for whom the belt was the only contributing factor). Where the choices differ is in the tricky cases (those where the belt was not the sole contributing factor).

Likewise if we had an effectiveness rating for crumple zones, the decision of whether to include crumple zones as a device when estimating the lives saved by seat belts, and if so, which attribution method to use, has ramifications for who will be said to be saved by belts when both belts and crumple zones play roles.

Although it is reasonable to think that, with millions of crashes occurring each year, there must be a fair number of crashes in which an occupant survived primarily because of some non-belt non-bag factor (such as the crumple zone, a guardrail, the side door beam, etc.) and to some lesser degree because of a seat belt and/or air bag, both choices imply that whenever a seat belt or air bag played a role in survival, only these factors should be attributed the survival of the occupants. This may at first seem disingenuous, but through examples such as the one we have illustrated above, one sees that there are potentially disagreeable aspects of any method.

Note that the treatment of air bags as a device or part of the setting also affects the potential fatalities and effectiveness ratings.

If we consider air bags to be part of the setting, then the potential fatalities are the people who die when we take away their seat belts. We will call these people the belt-potential fatalities for the purpose of this subsection. These people might live if they buckled up, or might die regardless.

If air bags are considered a device, and not part of the setting, then the potential fatalities, which we will call the belt/bag-potential fatalities for this subsection, are those people who would die unbelted and without an air bag.

The Belt-potential fatalities comprise a subset of the belt/bag-potential fatalities. The belt-potential fatalities experience, on average, more severe instances of the setting. That is, the typical crash in which one would die unbelted is more severe than the typical crash in which one would die unbelted and without an air bag.

The notion of belt-potential fatalities depends on the extent to which air bags are in vehicles. The typical crash in which one would die unbelted has been getting more severe in recent years, as air bag prevalence has increased.

Note also that the belt-potential fatalities and the bag-potential fatalities (i.e., the potential fatalities when one considers air bags as the sole device) overlap, and the belt/bag-potential fatalities strictly contain their union. For instance, people who would live if they were either belted or had air bags, but would die with neither, are a belt/bag-potential fatalities, but neither belt-potential fatalities nor a bag-potential fatalities.

Figure 22: The Effect on the Potential Fatalities of Considering Frontal Air Bags to Be a Device Versus Part of the Setting

The entire disk represents the potential fatalities when seat belts and air bags are both considered devices.

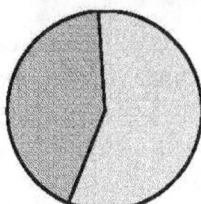

Considering seat belts as the sole device (and air bags to be part of the setting) results in a smaller set of potential fatalities.

Source: NCSA, NHTSA

If air bags were in every vehicle, then the effectiveness of seat belts that would result from the one-device consideration is the residual effectiveness $e_{belt|bag}$ from the two-device point of view.

Since, however, air bags do not currently protect all occupants, then the effectiveness of seat belts that would result from the one-device consideration is dependent on the extent to which air bags are present. The one-device effectiveness of seat belts decreases as the air bag prevalence increases. If no one had air bags, it would be the same as the two-device effectiveness of seat belts. If everyone had air bags, it would be the residual effectiveness of seat belts (conditioned on air bags).

A1.4 Summary: How Should One Choose the Devices?

The choice of whether to treat factors as devices or part of the setting can be quite complex, with substantial consequences. This choice affects both who will be said to have been saved, and what devices saved the survivors.

For certain factors, the choice is easy. The factors whose benefits you are interested in estimating must be treated as devices. For example, since this report concerns the lives saved by seat belts and frontal air bags, these had to be treated as devices in our calculation.

All devices must be rated for effectiveness. One must have the fatality-reducing effectiveness of all feasible combinations of devices in order to compute the lives saved by all devices combined. And depending on the method one chooses to attribute survivors to devices individually, one might need additional ratings, such as the rating of each device in the absence of all other devices.

In the case of seat belts and air bags, as we mentioned, we could not consider amorphous factors such as weather and road conditions as devices because we do not have effectiveness ratings for them. However we could have considered as devices any of the technologies implemented as a result of the FMVSS, such as front-disk brakes. Such technologies were all rated for effectiveness (for crashes in vehicles containing all previous FMVSS), and so we could have included these as devices.

Including devices in addition to seat belts and air bags would have resulted in different, but equally palatable, assessments of who was saved and by what. We choose to limit the devices to seat belts and air bags to be consistent with previous NHTSA calculations of the lives saved by these technologies.

NHTSA has used this choice of devices when calculating the lives saved by seat belts and air bags since bags were introduced in the calculations in 1985. Prior to 1985 and the appearance of appreciable numbers of air bags in vehicles, the agency considered seat belts to be the sole devices.

References

1.   Blincoe, L. (1994, June). *Estimating the Benefits from Increased Safety Belt Use*, Technical Report, DOT HS 808 133. Washington, DC: National Highway Traffic Safety Administration.

2.   Blincoe, L., et al. (2002, May). *The Economic Impact of Motor Vehicle Crashes, 2000*, Technical Report. DOT HS 809 446, May 2002. Washington, DC: National Highway Traffic Safety Administration.

3.   Federal Motor Vehicle Safety Standard 208. (1996, October). *Occupant Crash Protection*, Code of Federal Regulations, Title 49, Volume 5, Part 571.

4.   NHTSA. (2001, November). *Fifth/Sixth Report to Congress, Effectiveness of Occupant Protection Systems and Their Use*. DOT HS 809 442. Washington, DC: National Highway Traffic Safety Administration.

5.   Glassbrenner, D. (2003, September). *Safety Belt Use in 2003*, Technical Report. DOT HS 809 646. Washington, DC: National Highway Traffic Safety Administration.

6.   Glassbrenner, D. (2008, May). *Seat Belt Use in 2007 – Use Rates in the States and Territories*, Research Note. DOT HS 810 949. Washington, DC: National Highway Traffic Safety Administration.

7.   Kahane, C. (2004, October). *Lives Saved by the Federal Motor Vehicle Safety Standards and Other Vehicle Safety Technologies, 1960-2002*. DOT HS 809 833. Washington, DC: National Highway Traffic Safety Administration.

8.   Kahane, C. (2000, December). *Fatality Reduction by Safety Belts for Front-Seat Occupants of Cars and Light Trucks*, Technical Report. DOT HS 809 199. Washington, DC: National Highway Traffic Safety Administration.

9.   Kahane, C. (1996, August). *Fatality Reduction by Air Bags*, Technical Report. DOT HS 808 470. Washington, DC: National Highway Traffic Safety Administration.

10.   Morgan, C. (1999, June). *Effectiveness of Lap/Shoulder Belts in the Back Outboard Seating Positions*, Technical Report. DOT HS 808 945. Washington, DC: National Highway Traffic Safety Administration.

11.   NCSA. (undated). *Traffic Safety Facts 2007 - Occupant Protection* Fact Sheet DOT 810 991. Washington, DC: National Highway Traffic Safety Administration.

12.   NCSA. (undated). *Traffic Safety Facts 2007 – Alcohol-Impaired Driving*, Fact Sheet. DOT 810 985. Washington, DC: National Highway Traffic Safety Administration.

13.   OMB. (2003, September). *Regulatory Analysis*, Circular A-4. Washington, DC: Office of Management and Budget.

14.   Starnes, M. (2005, March). *Lives Saved Calculations for Infants and Toddlers*, Research Note. DOT HS 809 778. Washington, DC: National Highway Traffic Safety Administration.

15.   Starnes, M. (2008, October). *Lives Saved in 2007 by Restraint Use and Minimum Drinking Age Laws*, Crash*Stat. DOT HS 811 049. Washington, DC: National Highway Traffic Safety Administration.

16.   Tessmer, J. (2008, April). *FARS Analytical Reference Guide 1975 to 2007*. Technical Report. DOT HS 810 937. Washington, DC: National Highway Traffic Safety Administration.

17.   Wang, J., & Blincoe, L. (1999, July). *The Estimated Air Bag Effect on Lives Saved by Belt Use*. Research Note. Washington, DC: National Highway Traffic Safety Administration.

18.   Wang, J., & Blincoe, L. (2001, June). *BELTUSE Regression Model Update*. Research Note. Washington, DC: National Highway Traffic Safety Administration.

19.   Wang, J., & Blincoe, L. (2003, May). *Belt Regression Model - 2003 Update*. Research Note, DOT HS 809 639. Washington, DC: National Highway Traffic Safety Administration.